Industry Water Conservation and Efficiency: Advancing Sustainability by Reducing Water Use and Supporting Recycling, Reuse, Supply Chains, and ESG Goals

Copyright

Industry Water Conservation and Efficiency: Advancing Sustainability by Reducing Water Use and Supporting Recycling, Reuse, Supply Chains, and ESG Goals

ISBN (eBook): 978-1-991370-98-3

ISBN (Paperback): 978-1-991370-99-0

Published by Our Future Water

First Edition, 2025

Cover design and interior layout by Our Future Water

Table of Contents

Introduction

The industrial sector is one of the world's largest consumers of freshwater, drawing on rivers, aquifers, and reservoirs to power economic growth and sustain global production systems. Industries ranging from energy and chemicals to textiles, food processing, and manufacturing rely on consistent water supplies for production, cooling, cleaning, and energy generation. However, this dependence places industry at the center of today's mounting water challenges. Population growth, rapid urbanization, and climate change are intensifying pressures on already scarce water resources. In many regions, overextraction, pollution, and competing demands from agriculture and cities compound the risks industries face.

Water is no longer an invisible input or low-cost utility—it has become a strategic resource that can define the future competitiveness of businesses. Governments, investors, and communities are increasingly scrutinizing how industries manage their water use and wastewater discharges. Companies that fail to address water risks may face regulatory penalties, reputational damage, or supply chain disruptions. Conversely, those that embed water conservation and efficiency into their operations can unlock resilience, cost savings, and social license to operate, while contributing to global sustainability agendas such as the United Nations Sustainable Development Goal 6 on clean water and sanitation.

Technological and managerial innovations are creating new pathways for industries to reduce their water footprints. Digital tools such as smart meters, sensors, and data platforms enable real-time monitoring, leak detection, and performance optimization. Process innovations, including advanced treatment systems and closed-loop cooling, help industries recycle and reuse water, reducing their reliance on freshwater withdrawals. These changes not only conserve water but often bring energy savings and emissions reductions, given the deep interconnections between water and energy systems.

Adopting efficiency and conservation strategies also positions industries to align with the circular economy, where waste streams are transformed into resources. For example, treated wastewater can be reused within the same facility or shared across industrial clusters, creating opportunities for symbiosis and shared infrastructure. Such systemic approaches illustrate that industrial water efficiency is not just about reducing consumption but also about reimagining how water is valued and integrated into business models.

This book provides a comprehensive exploration of how industries can transition toward sustainable water use. It covers the global context of industrial water demand, the drivers pushing businesses toward conservation, the technologies and practices available for improving efficiency, and the governance and financing models that enable change. It also looks ahead to future pathways, highlighting emerging innovations and global strategies that can scale solutions across sectors and regions.

In doing so, the book emphasizes that water conservation and efficiency are essential for building industrial resilience in a climate-impacted, resource-constrained world. By embedding water sustainability into strategies and operations, industries not only mitigate risks but also contribute to broader environmental protection and long-term economic viability.

Chapter 1: Global Context of Industrial Water Use

Industries around the world are under increasing pressure to manage water resources responsibly as demand rises and availability becomes more uncertain. From manufacturing and energy production to mining and food processing, industrial activities account for a significant share of global freshwater withdrawals. This chapter explores the broad context in which industries operate, focusing on the trends, drivers, and challenges shaping water use at a global scale. By understanding these dynamics, industries can better position themselves to adapt, innovate, and contribute to long-term water security while maintaining competitiveness in a rapidly evolving landscape.

Industrial Water Demand and Trends

Industrial water demand has grown rapidly over the past century, driven by the expansion of manufacturing, energy production, and resource-intensive industries. In many economies, industry is now the second-largest consumer of freshwater after agriculture, accounting for a significant share of withdrawals worldwide. While the proportion of water used by industry varies across regions, the trend is clear: as countries industrialize, their water demand shifts from predominantly agricultural use to a growing share allocated to factories, processing plants, and power generation facilities. This transition reflects both the scale of modern industrial systems and the complexity of processes that require substantial volumes of water.

The use of water in industry is diverse. It is essential for cooling in thermoelectric power plants, a critical input in chemical reactions, a carrier in pulp and paper production, a solvent in textiles and pharmaceuticals, and a key ingredient in food and beverage processing. Beyond direct use, water is embedded throughout supply chains, linking industrial production to agricultural inputs and raw material extraction. As a result, the industrial water footprint is far larger than what is withdrawn on-site, highlighting the importance of managing both direct and indirect water use.

Global water demand from industry is expected to continue rising, particularly in emerging economies where industrialization is accelerating. Countries in Asia, Latin America, and parts of Africa are expanding their manufacturing and energy sectors to support economic growth and rising populations. This expansion increases pressure on water resources that are often already stressed by agricultural demand and urban consumption. At the same time, industries in developed economies face the challenge of modernizing aging infrastructure and processes to reduce water intensity and comply with stricter environmental standards.

Trends in industrial water use also reflect technological and regulatory shifts. Many industries are adopting more efficient equipment, closed-loop cooling systems, and advanced wastewater treatment technologies to reduce overall consumption. The drive toward sustainability has encouraged companies to monitor and report their water performance, creating benchmarks and competitive incentives for conservation. However, despite these advances, global industrial water demand is projected to rise as economic development and energy transitions continue. The International Energy Agency notes that energy production alone could significantly increase water withdrawals if new technologies are not coupled with efficiency measures.

Climate change adds another layer of complexity to industrial water demand. More frequent droughts, variable rainfall, and declining snowpack in many regions are altering water availability. Industries dependent on reliable supplies, such as hydropower, thermal power, and large-scale manufacturing, face disruptions when water sources become unreliable. This volatility highlights the importance of long-term planning and resilience strategies to ensure continuity of operations.

Consumer and investor expectations are also shaping trends. As public awareness of water scarcity grows, industries are being asked to demonstrate accountability and stewardship. Disclosure initiatives and sustainability reporting frameworks require companies to quantify and reduce their water footprints. Failure to do so can affect brand reputation and investor confidence, while proactive management can enhance market positioning and stakeholder trust.

Looking forward, industrial water demand and trends will be shaped by three main forces: economic development, technological innovation, and climate resilience. Emerging economies will continue to drive absolute increases in demand, while advanced economies may stabilize or reduce withdrawals through efficiency gains. Innovations in digital monitoring, water reuse, and industrial symbiosis will create opportunities to decouple growth from water use. Meanwhile, climate-related risks will force industries to integrate water conservation into core business planning.

Together, these trends highlight the dual challenge and opportunity facing industries. Water is a finite, shared resource under growing stress, yet it is also an area where innovation, collaboration, and efficiency can deliver both business resilience and societal benefits. Industry's response to these trends will play a central role in shaping global water security in the decades ahead.

Regional Variations in Water Stress

Water stress is not evenly distributed across the globe, and the industrial sector experiences these regional variations in profound ways. While some countries enjoy abundant water resources relative to their population and industrial base, others face severe scarcity that constrains both current operations and future growth. The regional picture is shaped by a combination of climatic conditions, economic development patterns, infrastructure capacity, and governance frameworks that together determine how water is allocated and managed.

In arid and semi-arid regions, such as the Middle East and North Africa, water scarcity is a structural condition. Industries in these areas often rely heavily on desalination, treated wastewater, or imported virtual water through trade. The high energy intensity of desalination and the limited reuse infrastructure create significant cost and sustainability challenges for industrial operators. As populations in these regions grow and energy transitions accelerate, industries face mounting pressure to reduce their reliance on freshwater withdrawals and adopt circular water strategies.

By contrast, water-rich regions such as parts of North America and Northern Europe have historically faced less stress, but this relative abundance does not eliminate risks. Climate change is altering precipitation patterns, leading to floods in some areas and droughts in others. For instance, industries in the western United States increasingly face restrictions during prolonged droughts, while those in northern Europe must manage water quality issues associated with agricultural runoff and urban expansion. Even in water-abundant regions, governance and infrastructure limitations can translate into local scarcity, particularly in areas where industries cluster around shared water sources.

Asia presents a particularly complex picture. On one hand, countries such as China and India are experiencing rapid industrial growth that drives rising water demand. On the other, many regions within these countries already face severe stress due to over-extraction of groundwater, pollution from untreated wastewater, and competition with agriculture and cities. Industrial zones in northern China, for example, have been forced to relocate or scale back operations because aquifers are being depleted faster than they can recharge. In South Asia, the reliance on seasonal monsoons makes water availability highly variable, exposing industries to periodic shortages that disrupt production.

Latin America, with its significant freshwater reserves in areas such as the Amazon Basin, is often viewed as water-secure, yet industrial water use there is uneven. While some regions have ample supplies, others—particularly along coastal and urbanized zones—face growing stress as industrial hubs compete with agriculture and cities. Additionally, pollution from extractive industries such as mining and oil production has created localized water quality crises, demonstrating that abundance at a national level does not guarantee security at a regional or local scale.

Africa faces unique challenges where industrialization is growing but infrastructure is often inadequate to meet demand. In many countries, weak water management systems, combined with climate variability, create hotspots of scarcity even in areas with relatively high rainfall. For industries seeking to expand in Africa, water security is both an operational and reputational issue, as communities expect companies to balance growth with equitable water use. The absence of reliable infrastructure often compels industries to invest in their own water supply and treatment systems, raising costs and reinforcing the importance of efficiency.

Regional variations in water stress also highlight the interconnected nature of global supply chains. A company headquartered in a water-abundant country may still face significant risks if its suppliers operate in water-scarce regions. For this reason, multinational corporations are increasingly adopting global water risk assessments, mapping stress across all operational and supply chain locations to identify vulnerabilities.

Ultimately, regional differences in water stress shape both the risks industries face and the strategies they adopt. Where scarcity is acute, industries must prioritize conservation, recycling, and alternative sources. Where abundance exists, the focus may be on protecting water quality and building resilience against climate variability. In all regions, the underlying message is the same: water cannot be taken for granted, and industries must adapt to the realities of their local context while recognizing the broader global dynamics that influence water security.

The Water-Energy Nexus in Industry

The interdependence between water and energy is one of the defining challenges for industrial sustainability. Every unit of energy produced or consumed in industry requires water somewhere along the chain, and every liter of water used in industrial processes typically requires energy for pumping, treatment, heating, or cooling. This relationship—known as the water-energy nexus—underscores how efficiency in one area can generate benefits in the other, while inefficiencies can compound resource stress and operational risks.

Industries that rely heavily on thermal energy production, such as power plants, illustrate the nexus most clearly. Water is indispensable for cooling systems, steam generation, and emissions control. Thermoelectric power plants account for some of the largest industrial withdrawals globally, and their dependence on freshwater creates vulnerabilities when rivers or reservoirs run low. Droughts and heatwaves have already forced power plants in several regions to reduce output or shut down, illustrating how water stress translates directly into energy insecurity. Similarly, water-intensive energy production methods such as biofuels, shale gas extraction, and hydropower all highlight the two-way relationship, where energy demand accelerates water demand and vice versa.

Within manufacturing and processing industries, the nexus is equally significant. Water is needed for cleaning, dissolving, and transporting raw materials, while energy is consumed to move, heat, and treat water. For example, in the food and beverage sector, both water and energy are essential for cooking, sterilization, and refrigeration. In textiles and chemicals, heating water for dyeing, finishing, and reaction processes can account for a major share of total energy use. Inefficiencies in water management, such as leaky distribution systems or poorly maintained cooling towers, thus carry a hidden energy cost, and the reverse is also true.

Climate change intensifies these interdependencies. Rising temperatures increase demand for cooling water in industrial processes and power plants, while also reducing the efficiency of existing cooling systems. At the same time, climate-driven variability in water supply reduces reliability for industries that require large, steady volumes of water. These dual pressures create a feedback loop: more energy is needed to secure, treat, and distribute diminishing water resources, while stressed energy systems demand more water, deepening vulnerabilities in regions already facing scarcity.

Addressing the water-energy nexus requires integrated solutions that recognize both dimensions simultaneously. Improvements in water efficiency—such as recycling, reuse, and advanced cooling technologies—can reduce energy requirements. Conversely, investments in energy efficiency, renewable energy, and low-carbon technologies often decrease water consumption by lowering reliance on water-intensive power generation. Digitalization strengthens this integration by enabling real-time monitoring of both water and energy flows, helping industries identify hotspots of inefficiency and optimize resource use across systems.

The financial implications of the nexus are considerable. Energy accounts for a significant share of the cost of water operations, particularly in pumping and treatment. In some industries, energy use for water management can exceed the direct cost of water itself. By reducing demand for one resource, companies often find cost savings in the other, creating opportunities for efficiency gains that directly improve profitability. This is particularly important in regions where water and energy tariffs are rising, making conservation measures financially attractive as well as environmentally responsible.

Recognizing and addressing the water-energy nexus is therefore central to industrial resilience. Industries that integrate water and energy efficiency into their operations not only reduce environmental impacts but also secure long-term operational stability. The nexus reinforces the need for systemic thinking: water and energy cannot be managed in isolation, and strategies that treat them as interconnected resources are essential for navigating the challenges of a climate-constrained, resource-limited world.

Impacts of Climate Change on Industrial Water Availability

Climate change is reshaping the global hydrological cycle in ways that directly affect industries reliant on secure, predictable water supplies. Rising temperatures, shifting precipitation patterns, glacial retreat, and more frequent extreme weather events are altering both the quantity and quality of available water. For industries, these changes translate into increased risks of operational disruptions, higher costs for water sourcing and treatment, and mounting pressure to adapt infrastructure and processes to new realities.

One of the most immediate impacts of climate change is the intensification of droughts in many regions. Prolonged dry periods reduce river flows, lower reservoir levels, and deplete aquifers, leaving industries without the volumes they require for production, cooling, or cleaning. Water-intensive sectors such as energy generation, mining, and chemicals are particularly vulnerable to these shortages. For example, thermal power plants may be forced to reduce output or shut down altogether when cooling water supplies fall below critical thresholds. Such events not only affect the industries themselves but also have ripple effects across economies that depend on reliable energy and industrial outputs.

At the other end of the spectrum, climate change is also increasing the frequency and severity of floods. Heavy rainfall and extreme storms can overwhelm drainage and treatment infrastructure, contaminating water sources with sediments, chemicals, or pathogens. Industries drawing on these sources may face sudden declines in water quality, requiring costly treatment upgrades or temporary suspension of operations. Flooding also damages physical assets, disrupting supply chains and raising insurance and recovery costs. In both drought and flood conditions, industries are learning that climate variability can no longer be treated as an occasional risk but must be factored into long-term planning.

Glacial retreat and declining snowpack represent another significant challenge. In many regions, glaciers and seasonal snowmelt provide essential water flows during dry months, buffering variability in precipitation. As these natural storage systems shrink, downstream industries lose a critical source of stability. The Andes, Himalayas, and Alps, for instance, support vast populations and industrial zones that depend on glacial runoff. Without it, industries face greater exposure to seasonal water shortages and heightened competition with agriculture and municipalities. This trend places a premium on efficiency and reuse, as natural water reliability continues to diminish.

Rising temperatures further complicate industrial water use by reducing the efficiency of cooling systems and increasing evaporation losses from reservoirs and storage facilities. In power generation, warmer river or lake water reduces cooling efficiency, forcing plants to use greater volumes to achieve the same output, just as supplies are becoming scarcer. Evaporation from industrial ponds, canals, and open storage tanks also increases under hotter conditions, raising the overall demand for water at a time when conservation is most needed.

Water quality is also at risk. Climate-driven changes in hydrology often lead to higher concentrations of pollutants in water bodies as flows decline, reducing dilution capacity. Algal blooms, which thrive in warmer, nutrient-rich waters, can contaminate supplies and complicate treatment processes for industries. In coastal regions, sea-level rise is accelerating saltwater intrusion into freshwater aquifers, particularly where over-extraction has already weakened natural defenses. Industries that depend on groundwater in such regions must contend with rising salinity, requiring more advanced and costly treatment to maintain water usability.

The economic and strategic implications of these changes are substantial. Industries operating in water-stressed regions face escalating costs for securing water rights, developing alternative supplies, or upgrading treatment infrastructure. Insurance premiums for facilities exposed to flood or drought risk are increasing, while investors are demanding greater disclosure of climate and water-related risks. Companies that cannot adapt may see declining competitiveness, while those that act early to build resilience may secure advantages in cost, reputation, and market access.

Ultimately, climate change underscores the need for industries to transition from reactive to proactive water management. Building resilience requires a combination of efficiency improvements, diversification of water sources, integration of reuse and recycling, and adoption of advanced monitoring and forecasting tools. Climate risks are no longer distant threats—they are already disrupting industrial water availability worldwide. The challenge for industry is to recognize water as a finite, climate-sensitive resource and embed adaptive strategies that ensure operations can withstand an uncertain hydrological future.

Chapter 2: Drivers of Water Conservation and Efficiency

Industrial water conservation and efficiency are shaped by a range of interlinked drivers that extend from regulatory mandates to economic imperatives and societal expectations. Understanding these drivers is essential for industries to recognize why water management has become a strategic priority and how external pressures influence corporate decisions. This chapter examines the factors pushing industries toward greater responsibility in water use, including policy frameworks, financial and economic pressures, stakeholder expectations, and alignment with global sustainability agendas. Together, these drivers illustrate the evolving landscape in which water is no longer just an operational concern but a determinant of long-term resilience and competitiveness.

Regulatory and Policy Frameworks

Regulatory and policy frameworks are among the most powerful drivers shaping how industries approach water conservation and efficiency. Around the world, governments are using laws, standards, and incentives to ensure that industries internalize the environmental and social costs of water use while encouraging innovation in efficiency. These frameworks vary widely across jurisdictions, reflecting differences in water availability, economic priorities, and institutional capacity, but they share a common objective: to align industrial water use with long-term sustainability and resilience.

At the foundation of these frameworks are water rights and allocation systems. In regions where water scarcity is acute, governments set strict withdrawal limits that industries must adhere to, often requiring permits or licenses tied to specific volumes and conditions. This approach places legal obligations on industries to monitor consumption, report usage, and invest in efficiency measures to stay within allocated quotas. In some cases, allocation frameworks are dynamic, adjusting entitlements based on annual hydrological conditions, which further motivates industries to build flexibility into their water strategies.

Pollution control regulations also play a crucial role in shaping industrial water use. Discharge standards, effluent limits, and water quality guidelines force industries to treat wastewater before releasing it into rivers, lakes, or aquifers. Compliance requires investment in treatment technologies and operational changes that often result in more efficient water use overall. Increasingly, regulations are moving beyond minimum compliance to encourage industries to recover and reuse water, fostering a shift from linear consumption to circular practices.

Economic instruments have become central features of regulatory frameworks. Pricing mechanisms, such as volumetric tariffs or tiered water rates, send price signals to industries to conserve water. Where subsidies for water supply once shielded companies from the true cost of their consumption, reforms are gradually moving toward cost-reflective tariffs that incorporate infrastructure, energy, and environmental costs. These economic tools make efficiency investments more attractive, as the payback period for water-saving technologies shortens under higher tariffs.

In parallel, governments are integrating water policy into broader environmental and climate strategies. National adaptation plans, sustainable development roadmaps, and carbon neutrality commitments frequently include water efficiency targets for industries. This integration highlights the recognition that water management is inseparable from climate resilience, biodiversity protection, and energy security. By embedding water into cross-cutting policy agendas, governments ensure that industries are accountable not only for their direct impacts but also for their contributions to broader sustainability outcomes.

Voluntary and incentive-based mechanisms complement regulatory approaches. Many governments establish certification schemes, labeling programs, or award systems that recognize industries achieving high levels of water efficiency. These programs encourage companies to go beyond compliance, creating reputational benefits and competitive advantages. Tax incentives, grants, and low-interest financing for water-efficient technologies further reinforce the business case for adopting conservation measures.

Policy frameworks also increasingly emphasize transparency and disclosure. Industries are being required to publish water use data, risk assessments, and sustainability reports. These measures align with international disclosure initiatives and investor expectations, ensuring that industrial water practices are visible to stakeholders. In turn, transparency fosters peer comparison and benchmarking, which motivates continuous improvement.

While regulatory and policy frameworks have made substantial progress in driving water conservation, challenges remain. Enforcement can be weak in regions with limited institutional capacity, and industries may resist reforms that increase operational costs. Moreover, fragmented governance across water, energy, and environmental agencies often leads to inconsistent or overlapping rules. Addressing these challenges requires integrated governance structures that bring together multiple stakeholders, ensuring coherence and clarity in industrial water management.

Overall, regulatory and policy frameworks are evolving from command-and-control models toward more dynamic, integrated approaches that combine regulations, economic incentives, and voluntary measures. For industries, this shift means adapting not only to compliance requirements but also to new expectations for leadership in water stewardship. By aligning business strategies with policy frameworks, industries can manage risks, improve efficiency, and contribute to the sustainable use of one of the planet's most vital resources.

Financial and Economic Pressures

Financial and economic pressures are among the strongest motivators pushing industries to adopt water conservation and efficiency measures. While environmental regulations and stakeholder expectations shape behavior, the economic realities of rising water costs, increasing energy expenses, and operational risks tied to water scarcity provide immediate, measurable incentives for businesses to act. Companies must balance the need for reliable water supplies with the imperative of maintaining profitability and competitiveness in a global marketplace where efficiency is increasingly synonymous with resilience.

The most direct pressure comes from the growing cost of water itself. In many regions, governments are moving toward cost-reflective pricing models that capture the full expense of sourcing, treating, and delivering water to industrial users. Subsidies that once masked these costs are being reduced or eliminated, exposing industries to higher tariffs that escalate as demand intensifies. For water-intensive sectors such as food processing, textiles, chemicals, and energy, these rising costs can represent a significant portion of operational expenses. Investing in efficiency, therefore, becomes a financial strategy to manage cost volatility and ensure long-term competitiveness.

Energy expenses are tightly linked to water use, reinforcing economic pressures. Pumping, treating, heating, and cooling water requires substantial amounts of energy, and inefficiencies in water use often translate directly into higher energy bills. As energy prices rise globally due to market fluctuations, carbon pricing, or transitions away from subsidized fossil fuels, industries are compelled to find ways to reduce both water and energy demand simultaneously. Improvements such as installing efficient pumps, adopting closed-loop cooling, or recycling wastewater not only conserve water but also cut energy expenditures, creating a dual financial incentive.

Water scarcity itself poses significant financial risks. When droughts or shortages force industries to curtail operations, the costs are immediate and severe, including lost production, supply chain disruptions, and penalties for non-delivery of goods or services. Insurance premiums for facilities in water-stressed regions are rising as underwriters recognize these risks, further adding to operational costs. Companies that fail to address vulnerabilities may face downgraded credit ratings or higher borrowing costs, as financial institutions increasingly incorporate water risk into their assessments of business resilience.

Global supply chains magnify these pressures. Even if a company operates in a water-abundant region, it may depend on suppliers located in water-scarce areas. Disruptions in these regions can create bottlenecks, increase procurement costs, or lead to reputational damage if suppliers are perceived to be mismanaging water resources. Multinational corporations are responding by conducting water risk assessments across their supply chains, pushing suppliers to adopt efficiency measures and in some cases shifting sourcing to reduce exposure. The financial implications of these strategies ripple across entire industries, reinforcing water efficiency as a condition for participation in global markets.

Investor expectations add another layer of economic pressure. Institutional investors and asset managers are increasingly demanding that companies disclose water risks and demonstrate strategies for managing them. Companies that fail to act may find themselves excluded from sustainability-focused investment funds or face declining shareholder confidence. Conversely, businesses that integrate water efficiency into their operations can attract investment by demonstrating resilience, cost savings, and alignment with environmental, social, and governance (ESG) criteria.

The reputational dimension of financial pressure cannot be overlooked. Public perception of industries that waste water or pollute local sources can lead to consumer boycotts, community opposition, and costly delays in project approvals. Social license to operate is increasingly tied to how responsibly companies manage shared resources, and poor performance can translate directly into financial setbacks. This reputational risk is particularly acute in water-scarce regions, where communities expect industries to act as responsible stewards.

Altogether, financial and economic pressures create a strong business case for water conservation and efficiency. Rising water and energy costs, exposure to scarcity risks, supply chain vulnerabilities, investor demands, and reputational considerations all converge to make efficiency a strategic necessity. By proactively addressing these pressures, industries not only secure cost savings but also position themselves as leaders in sustainability, capable of navigating a future where water is recognized as both a vital resource and a defining factor of economic resilience.

Stakeholder and Community Expectations

Stakeholder and community expectations have become decisive forces in shaping how industries approach water conservation and efficiency. While regulatory frameworks and financial pressures set formal boundaries for action, the demands of local communities, employees, consumers, investors, and advocacy groups create social and reputational pressures that can be just as powerful. In a world where transparency and accountability are increasingly valued, industries must demonstrate that their operations not only comply with laws but also contribute to the well-being of people and ecosystems that depend on shared water resources.

Communities located near industrial facilities are often the most vocal stakeholders. They depend on the same rivers, lakes, and aquifers for drinking water, agriculture, and cultural uses, and they quickly feel the impacts of overuse or pollution. When industrial withdrawals reduce local water availability, tensions can escalate, leading to protests, legal disputes, or delays in project approvals. Similarly, untreated or poorly managed wastewater discharges that compromise water quality can spark strong opposition. For industries, earning and maintaining community trust requires visible commitment to conservation, efficiency, and environmental protection. Companies that fail to respond to community concerns risk losing their social license to operate, a critical factor in sustaining long-term business continuity.

Employees are another key stakeholder group shaping expectations. Many workers, particularly in younger generations, prefer to be associated with companies that take sustainability seriously. A strong commitment to water conservation and efficiency can improve employee morale, enhance recruitment, and reduce turnover. Conversely, perceptions of negligence or disregard for environmental responsibility can harm a company's ability to attract skilled labor. Embedding water stewardship into corporate culture and operations demonstrates to employees that the company is aligned with broader societal values.

Consumers also exert influence through their purchasing choices and brand perceptions. In sectors where water is integral to production—such as beverages, textiles, and electronics—public awareness campaigns and advocacy initiatives have raised consumer expectations for responsible water use. Labels, certifications, and corporate sustainability reports allow customers to evaluate companies on their environmental performance. Businesses that demonstrate leadership in water conservation can differentiate themselves in competitive markets, while those perceived as wasteful or irresponsible risk reputational damage that translates into declining sales or weakened brand loyalty.

Investors and financial institutions represent another critical stakeholder group. Increasingly, they view water as a material risk that can affect profitability, supply chain resilience, and long-term value. Investor coalitions and disclosure initiatives, such as those aligned with ESG frameworks, require companies to assess and report their water risks. Firms that fail to provide credible strategies for managing water use may face higher costs of capital or exclusion from investment portfolios. In contrast, those that meet investor expectations by integrating efficiency into their operations and disclosing progress transparently are better positioned to attract funding and enhance shareholder confidence.

Civil society organizations, including non-governmental organizations (NGOs) and advocacy groups, play a further role in shaping expectations. These groups monitor industrial practices, raise awareness of water issues, and campaign for stronger protections. Their influence can extend to shaping public opinion, lobbying policymakers, and mobilizing communities. Industries that proactively collaborate with NGOs or engage in multi-stakeholder platforms often find opportunities to co-create solutions, enhance credibility, and avoid conflict. On the other hand, adversarial relationships with advocacy groups can generate negative publicity and long-lasting reputational harm.

The expectations of stakeholders and communities are not static; they evolve with growing awareness of global water challenges and climate impacts. As scarcity intensifies and competition for resources increases, tolerance for inefficient or polluting industrial practices diminishes. Industries are therefore under continuous pressure to improve transparency, adopt advanced efficiency measures, and demonstrate tangible contributions to sustainable water management.

Meeting these expectations requires a proactive approach. Companies must engage stakeholders early and consistently, communicate openly about risks and strategies, and provide evidence of measurable progress. This involves not only technical investments in water efficiency but also cultural shifts within organizations that prioritize water stewardship as a core business value. By aligning operations with stakeholder and community expectations, industries can strengthen trust, reduce conflict, and build reputational capital that supports long-term resilience. Ultimately, industries that embrace these expectations will be better positioned to thrive in a world where water is recognized as both a shared resource and a shared responsibility.

Alignment with Global Sustainability Agendas

Industrial water conservation and efficiency are no longer viewed as isolated operational practices but as essential components of global sustainability agendas. From the United Nations Sustainable Development Goals (SDGs) to climate agreements and regional environmental frameworks, industries are expected to align their water use with broader objectives that balance economic growth, environmental stewardship, and social equity. This alignment reflects the growing recognition that sustainable water management is fundamental to achieving long-term global development and resilience.

The UN SDGs provide the most widely recognized framework for linking industrial practices to global priorities. SDG 6, which calls for the availability and sustainable management of water and sanitation for all, directly addresses industrial water use by emphasizing efficiency, pollution reduction, and integrated water resources management. Beyond SDG 6, industrial water practices contribute to other goals as well: SDG 9 on industry, innovation, and infrastructure encourages sustainable industrialization; SDG 12 on responsible consumption and production highlights efficient resource use; and SDG 13 on climate action links water conservation with adaptation and mitigation strategies. By embedding water efficiency into operations, industries contribute simultaneously to multiple interconnected goals.

International climate frameworks also reinforce the importance of industrial water efficiency. The Paris Agreement emphasizes the need for climate adaptation, and water is central to this agenda. Industries that adopt water-efficient processes reduce vulnerability to climate-induced scarcity while also lowering their carbon footprint by minimizing energy-intensive water treatment and pumping. The water-energy nexus, a recurring theme in global discussions, highlights how conserving water helps reduce greenhouse gas emissions, aligning industrial strategies with climate commitments made by governments and businesses alike.

Regional frameworks and transboundary agreements further shape expectations for industries. In the European Union, for example, the

Water Framework Directive requires member states to achieve "good status" for water bodies, placing obligations on industries to reduce consumption and pollution. In North America, sustainability reporting standards and water stewardship initiatives have become embedded in corporate governance practices. In regions such as Africa and Asia, regional strategies emphasize industrial water efficiency as a pathway to economic development that does not compromise ecological integrity or community needs. These frameworks highlight that global agendas are interpreted and implemented locally, but industries everywhere are expected to contribute to shared objectives.

Corporate sustainability commitments also increasingly align with global agendas. Many multinational companies have pledged to achieve water neutrality, zero liquid discharge, or substantial reductions in water intensity as part of their broader ESG strategies. These commitments often reference global frameworks, signaling to investors and stakeholders that the company's water practices are aligned with recognized international standards. Reporting platforms such as the Global Reporting Initiative (GRI) and the CDP Water Security program provide tools for industries to disclose performance in ways that are comparable and credible, reinforcing the link between corporate action and global agendas.

Alignment with sustainability agendas also brings competitive advantages. Companies that position themselves as leaders in water stewardship can access new markets, secure investment, and strengthen relationships with governments and communities. Participation in global initiatives, such as the CEO Water Mandate or the Alliance for Water Stewardship, allows industries to demonstrate leadership, share best practices, and influence the evolution of sustainability standards. These opportunities go beyond compliance, positioning companies as contributors to systemic change rather than passive participants.

However, alignment also demands accountability. Global agendas are increasingly accompanied by expectations for measurable progress, transparent reporting, and independent verification.

Industries must move beyond aspirational statements to implement tangible actions that can be tracked and evaluated against global benchmarks. This shift requires investment in data collection, performance monitoring, and disclosure systems, ensuring that industrial contributions to sustainability are both real and visible.

In sum, aligning industrial water conservation and efficiency with global sustainability agendas is both a responsibility and an opportunity. It ensures that industries are not only reducing risks and costs but also contributing to international efforts to safeguard ecosystems, mitigate climate impacts, and promote equitable development. By embedding these agendas into their strategies, industries can position themselves as active partners in building a more resilient and sustainable world, where water efficiency is recognized as a cornerstone of global progress.

Chapter 3: Industrial Water Audits and Benchmarking

Industrial water audits and benchmarking provide the foundation for identifying inefficiencies, measuring performance, and guiding continuous improvement in water management. While many companies recognize the importance of reducing water use, they often lack the structured approaches necessary to pinpoint where the greatest savings can be made. Audits serve as diagnostic tools, enabling industries to map consumption patterns, identify hotspots, and highlight opportunities for efficiency. Benchmarking, meanwhile, allows industries to compare performance across facilities, sectors, or regions, creating baselines that support target-setting and accountability. This chapter examines how audits and benchmarking together establish the evidence base required to drive transformative action in industrial water conservation.

Conducting Water Audits

Water audits are foundational tools for industries seeking to improve conservation and efficiency. Much like an energy audit identifies areas of waste and opportunities for savings, a water audit provides a systematic examination of how water is sourced, used, reused, and discharged within industrial operations. By mapping these flows, companies can quantify their water footprint, identify inefficiencies, and design targeted strategies for reducing consumption and costs. Conducting water audits has become an essential first step for industries that aim to transition from reactive water management to proactive stewardship.

A comprehensive water audit begins with data collection. Industries must compile records of water use from various sources, including municipal supply bills, groundwater extraction permits, and on-site measurements from meters or sensors. This data provides a baseline understanding of total consumption and helps distinguish between different sources of water. In many cases, industries find that their records are incomplete or fragmented, highlighting the importance of installing accurate metering systems to capture reliable data. Without robust baseline information, opportunities for efficiency may remain hidden or underestimated.

Once data is collected, the next step is to map water flows throughout the facility. This involves documenting how water enters the site, moves through processes, and leaves as effluent, steam, or product content. Flow diagrams are often used to visualize the pathways, making it easier to see where the largest volumes are consumed or lost. For example, in a textile factory, water may flow through dyeing, washing, and finishing processes before being discharged. Each stage can then be assessed for inefficiencies, such as excessive rinse cycles, leaks, or outdated equipment.

Identifying "hotspots" is one of the most valuable outcomes of a water audit. Hotspots are areas within the facility where water consumption is disproportionately high or where significant losses occur. Common hotspots include cooling systems, cleaning operations, and leak-prone distribution networks. Pinpointing these areas allows managers to prioritize interventions with the greatest potential impact. In many cases, simple measures such as repairing leaks, optimizing cleaning schedules, or retrofitting cooling towers yield immediate savings without large capital investments.

Beyond physical flows, water audits also consider quality. The quality of water required for different processes can vary significantly, and industries often use high-quality freshwater where lower-quality water could suffice. For instance, potable-grade water may be used for cooling or cleaning when treated wastewater could perform the same function. Audits help identify mismatches between water quality and application, creating opportunities to substitute sources and reduce dependence on scarce freshwater supplies.

Water audits also integrate financial analysis. By linking consumption data to costs, industries can calculate the economic value of water savings. This includes not only the direct cost of water purchases but also associated energy, treatment, and disposal expenses. In many cases, the hidden costs of water use—such as energy for pumping or chemical treatment for effluent—exceed the cost of the water itself. By quantifying these costs, audits provide a compelling business case for efficiency measures, often revealing short payback periods for investments in conservation technologies.

Digital tools are transforming how water audits are conducted. Advanced metering infrastructure, sensors, and data analytics platforms allow industries to track water use in real time, identify anomalies, and generate detailed performance reports. These technologies enhance the accuracy of audits and enable continuous monitoring rather than one-time assessments. For industries operating in water-stressed regions or under strict regulatory regimes, such digital capabilities are becoming essential for compliance, risk management, and operational resilience.

Stakeholder engagement is another critical dimension of water audits. Involving employees, suppliers, and local communities in the audit process can generate valuable insights and strengthen trust. Workers on the factory floor often have firsthand knowledge of where water is wasted, while suppliers can provide information on embedded water in raw materials. Communities, in turn, may highlight external impacts of industrial water use, such as reduced availability for agriculture or household needs. Incorporating these perspectives ensures that audits capture the full scope of risks and opportunities.

The outcomes of a water audit are most effective when integrated into a continuous improvement cycle. This means setting clear efficiency targets, implementing measures, monitoring progress, and conducting follow-up audits to track performance over time. Companies that institutionalize this cycle move beyond compliance-driven approaches to embed water efficiency as a core operational value. Over time, this fosters a culture of conservation that permeates decision-making at all levels of the organization.

In conclusion, conducting water audits equips industries with the knowledge needed to reduce consumption, lower costs, and mitigate risks. By providing a detailed picture of water flows, hotspots, and financial implications, audits transform water management from guesswork into strategic action. As pressures on global water resources intensify, industries that regularly conduct thorough audits will be better prepared to adapt, innovate, and thrive in an increasingly resource-constrained world.

Benchmarking Water Performance Across Sectors

Benchmarking water performance across sectors is a vital practice that enables industries to evaluate how efficiently they use water compared to peers, competitors, and global best practices. It provides a reference point for assessing progress, setting goals, and identifying areas where improvements can be made. By comparing performance metrics across different industries and geographies, companies can learn from leaders, adapt proven strategies, and avoid inefficiencies that may be common within their sector. Benchmarking transforms water conservation from an internal exercise into a broader competitive and collaborative effort.

The process begins with establishing metrics that capture water use and efficiency in standardized ways. Common indicators include total water withdrawals, water intensity (water used per unit of output), and the proportion of water recycled or reused. For example, in the beverage industry, water intensity may be measured in liters of water used per liter of product, while in energy generation, the benchmark could be cubic meters of water per megawatt-hour of electricity. These sector-specific metrics make it possible to evaluate performance in meaningful terms and account for the unique processes and demands of each industry.

Benchmarking highlights significant variations between sectors. Water-intensive industries such as textiles, pulp and paper, mining, and chemicals often exhibit higher water footprints per unit of output than less water-intensive sectors like electronics or pharmaceuticals.

Even within a single sector, performance can vary widely depending on geographic location, technology adoption, and regulatory environments. A textile factory in a water-scarce region that has invested in advanced dyeing technologies may achieve much lower water intensity than a similar facility in a water-abundant region relying on older equipment. Such contrasts underscore the importance of using benchmarking not only to measure averages but also to spotlight leaders whose practices can guide broader sectoral improvements.

Global and regional initiatives play a significant role in creating benchmarking platforms. Organizations such as the World Business Council for Sustainable Development (WBCSD), the Carbon Disclosure Project (CDP), and the Alliance for Water Stewardship (AWS) encourage companies to disclose water use data, which can then be aggregated and compared across sectors. These efforts provide industries with insights into where they stand relative to global peers, helping them identify gaps and opportunities. Benchmarking data also supports policymakers and regulators in setting realistic but ambitious efficiency standards that reflect proven industry practices.

Benchmarking is not only about measuring efficiency but also about driving innovation. By identifying top performers within or across sectors, benchmarking exercises reveal the technologies, management approaches, and cultural factors that enable superior performance. For instance, industries that excel in water recycling may rely on advanced membrane filtration or closed-loop systems that can be replicated elsewhere. Benchmarking thus creates pathways for knowledge transfer, enabling lagging companies to adopt proven solutions and accelerate progress.

Investors and financial institutions also use benchmarking to evaluate corporate water performance. With growing emphasis on ESG criteria, companies that benchmark well compared to peers may gain preferential access to financing or inclusion in sustainability-focused investment portfolios. Conversely, poor performance relative to sectoral benchmarks can raise red flags about long-term resilience, prompting investors to demand corrective action or reconsider support. This financial dimension adds another layer of incentive for companies to engage in benchmarking.

One of the challenges in benchmarking is ensuring consistency and comparability of data. Industries often operate under different reporting standards, measurement methods, or disclosure requirements. Water availability and quality also vary significantly by region, making it difficult to compare performance across geographies. To address these challenges, there is growing momentum toward harmonized reporting frameworks, such as the GRI standards or CDP questionnaires, which provide common definitions and methodologies. Such harmonization is essential for creating reliable and actionable benchmarks.

Ultimately, benchmarking water performance across sectors creates transparency, accountability, and motivation for continuous improvement. It helps companies understand their position in a competitive landscape, learn from leaders, and identify cost-effective strategies to reduce consumption and risk. As global water challenges intensify, benchmarking will remain a critical mechanism for industries to measure, compare, and elevate their performance, ensuring that water efficiency becomes a shared goal across all sectors.

Identifying Hotspots and Inefficiencies

Identifying hotspots and inefficiencies is a critical step in industrial water management because it highlights the points in a system where the greatest opportunities for savings and conservation exist. Hotspots are areas of disproportionate water use or loss, and inefficiencies are the processes, practices, or infrastructure that allow water to be wasted. By systematically identifying and addressing these problem areas, industries can reduce costs, strengthen resilience, and improve sustainability performance.

Hotspots typically emerge in parts of industrial operations where water is used most intensively or where processes are less efficient. Cooling systems are often a prime example, as they can account for a significant share of total industrial water withdrawals. In facilities that rely on once-through cooling, large volumes of water pass through the system and are discharged without reuse, representing an inefficiency that can be corrected by upgrading to closed-loop systems. Similarly, cleaning and sanitation processes can become hotspots if they rely on outdated practices, excessive rinse cycles, or high-quality water where lower grades would suffice.

Another common source of inefficiency is leakage within distribution networks. Poorly maintained pipes, valves, and storage tanks can result in substantial water loss before it even reaches the point of use. Identifying leaks often requires detailed monitoring and regular maintenance, as they may not be visible to operators but still contribute to higher costs and unnecessary withdrawals. Modern technologies such as acoustic sensors and smart metering can detect anomalies quickly, helping industries address leaks before they escalate.

Energy-intensive processes often reveal inefficiencies that are linked to water use. For example, heating water for industrial processes consumes both water and energy. If heat recovery systems are not in place, significant resources may be wasted. The water-energy nexus is central to identifying inefficiencies because water waste often implies energy waste, and vice versa. Pinpointing these overlaps allows industries to pursue measures that deliver combined savings.

Hotspots also arise where water quality is mismatched with its application. Industries frequently use potable-grade water in processes where reclaimed or lower-quality water would be sufficient. For example, cooling towers or cleaning operations do not always require high-quality inputs. Identifying these mismatches allows companies to reduce dependence on freshwater and increase reliance on recycled or alternative sources, improving both efficiency and sustainability.

Mapping water use across processes is the most effective way to identify hotspots and inefficiencies. By developing flow diagrams and quantifying water use at each stage of production, industries can visualize where the greatest withdrawals or losses occur. Benchmarking against sectoral averages or best-in-class performers can further highlight inefficiencies, showing where a facility lags behind peers. This process often reveals quick wins—such as replacing nozzles, optimizing rinse cycles, or repairing leaks—as well as longer-term investment opportunities in advanced treatment or recycling systems.

Financial analysis is another essential tool for identifying inefficiencies. By linking water consumption to direct and indirect costs, industries can uncover the hidden financial impacts of waste. For instance, every unit of water wasted carries with it the cost of sourcing, pumping, heating, and disposal. When these costs are fully accounted for, even small inefficiencies can be shown to have significant financial consequences. This perspective strengthens the business case for investing in conservation technologies and practices.

In addition to internal assessments, stakeholder input can provide valuable insights into hotspots and inefficiencies. Employees often recognize day-to-day practices where water is wasted, while local communities may highlight broader impacts such as reduced river flows or degraded water quality. Engaging stakeholders ensures that inefficiency identification is not limited to technical audits but also reflects operational realities and external perceptions.

Overall, identifying hotspots and inefficiencies transforms water management from a broad sustainability goal into a targeted strategy with tangible outcomes. By focusing on areas of greatest use, loss, or mismatch, industries can maximize the impact of efficiency initiatives, reduce operational risks, and enhance long-term resilience. The process requires accurate data, careful analysis, and a willingness to act on findings, but the rewards—in terms of cost savings, risk reduction, and sustainability performance—are substantial.

Linking Audits to Continuous Improvement

Water audits achieve their full potential when they are tied to continuous improvement rather than treated as isolated exercises. A single audit provides a snapshot of water use and highlights opportunities for efficiency, but without follow-up, monitoring, and adaptation, those opportunities may fade or remain underutilized. Continuous improvement ensures that water management becomes a dynamic process where lessons learned from each audit inform future strategies and embed conservation into the culture of the organization.

The first link between audits and continuous improvement lies in establishing baselines and setting measurable targets. Audit data provides the reference point for understanding how much water is used, where it is consumed, and where losses occur. From this baseline, companies can define realistic and time-bound goals for reduction, recycling, or substitution. When these goals are incorporated into broader sustainability strategies, they create accountability while also aligning with regulatory requirements and global environmental commitments.

Monitoring progress is equally essential. After implementing measures identified through audits, industries need systems to track ongoing performance. Smart meters, flow sensors, and digital dashboards make it possible to measure consumption in real time, compare outcomes against established benchmarks, and detect inefficiencies as they emerge. This feedback loop allows for timely course corrections and ensures that gains are sustained rather than eroded over time.

Employee involvement strengthens the continuous improvement cycle. Workers closest to daily operations often notice waste that is invisible in audit reports. Training programs, awareness campaigns, and recognition schemes encourage staff to contribute ideas and reinforce a shared responsibility for conservation. When employees see that audit findings translate into practical improvements, they are more likely to engage actively in identifying and sustaining new efficiencies.

The financial aspect also supports ongoing progress. Linking audit results to cost savings creates a strong business case for reinvesting those savings into additional efficiency measures. Reduced water bills, lower energy use, and smaller wastewater treatment costs can be quantified and tracked, demonstrating to management and investors that efficiency is not just environmentally responsible but also profitable. Over time, this reinvestment cycle can fund more advanced technologies and infrastructure upgrades.

Regular audits keep the process dynamic and responsive. Facilities evolve as new technologies, processes, or products are introduced, and water use patterns often shift as a result. Climate variability and regulatory changes also alter the external environment. Conducting audits at regular intervals ensures that new risks and opportunities are identified promptly, maintaining alignment between operational practices and emerging challenges.

External benchmarking and certification frameworks enhance this cycle by creating standards for comparison and accountability. Participation in programs such as the Alliance for Water Stewardship or reporting through established disclosure platforms pushes companies to demonstrate progress consistently and transparently. These mechanisms encourage organizations to maintain momentum, while also learning from peers and industry leaders.

Linking audits to continuous improvement, therefore, creates an iterative process where each assessment informs new goals, monitoring reveals performance, employees contribute solutions, and outcomes feed back into future planning. This approach transforms water audits from static reports into living tools that guide adaptation and resilience. Industries that embed this cycle into their operations are better equipped to achieve lasting reductions in water use, strengthen trust with stakeholders, and position themselves for long-term sustainability in a resource-constrained world.

Chapter 4: Technological Innovations for Water Efficiency

Technological innovations are redefining how industries approach water conservation and efficiency, providing new tools to address challenges that traditional methods alone cannot solve. As demand grows and freshwater availability declines, companies are turning to advanced technologies to monitor use, optimize processes, recycle wastewater, and minimize losses. This chapter explores the range of solutions now available, from smart metering and digital twins to membrane filtration, advanced cooling, and zero-liquid discharge systems. By adopting these innovations, industries can achieve significant gains in efficiency, reduce costs, and align operations with sustainability goals, positioning themselves for resilience in a resource-constrained future.

Advanced Monitoring and Smart Metering

Advanced monitoring and smart metering have emerged as transformative tools for industrial water management, enabling companies to track consumption with precision, identify inefficiencies in real time, and optimize operations to reduce waste. Traditional methods of water monitoring often relied on periodic manual readings of meters or estimations based on utility bills, which provided only limited insights and delayed feedback. In contrast, advanced systems use digital technologies, sensors, and data analytics to deliver continuous, granular information about water flows, quality, and usage patterns. This shift empowers industries to move from reactive responses to proactive and predictive water management strategies.

At the core of this transformation are smart meters—devices that record water use at high frequency intervals and transmit data to centralized systems. Unlike traditional meters, smart meters capture detailed consumption patterns across different times of day, production cycles, or facility zones. This granularity allows industries to distinguish between baseline use, peak demand, and irregular consumption, helping pinpoint inefficiencies such as leaks, excessive rinsing, or equipment malfunctions. For instance, a spike in water use outside production hours could indicate a hidden leak or faulty valve that would otherwise go unnoticed.

Smart metering is often complemented by advanced monitoring systems that use a network of sensors and Internet of Things (IoT) devices. These sensors can track variables such as flow rate, pressure, temperature, and water quality parameters including pH, turbidity, and salinity. By integrating this data, companies gain a comprehensive view of how water moves through their facilities and how it is being used in different processes. Monitoring water quality is particularly valuable in industries that require specific standards, such as food and beverage, pharmaceuticals, or electronics manufacturing. Real-time alerts ensure that any deviations from required quality levels are detected immediately, reducing the risk of product defects or regulatory violations.

The data generated by smart meters and sensors is most powerful when combined with analytics platforms that provide visualization, benchmarking, and predictive capabilities. Dashboards can display water use in real time across different departments, allowing managers to compare performance and identify hotspots. Predictive analytics can forecast future consumption based on production schedules, seasonal variations, or climate conditions, enabling companies to plan resource use more effectively. Machine learning algorithms can even identify usage patterns that humans might overlook, suggesting targeted interventions to optimize processes.

Beyond operational efficiency, advanced monitoring and smart metering support regulatory compliance and sustainability reporting. Many jurisdictions require industries to submit detailed records of water withdrawals, discharges, and quality parameters. Automated monitoring systems simplify this process, ensuring accurate and timely data collection while reducing the administrative burden. For companies participating in voluntary disclosure programs or ESG reporting, smart metering provides the transparency and credibility needed to demonstrate leadership in water stewardship.

Financial benefits are another major driver of adoption. By identifying leaks and inefficiencies quickly, smart systems reduce water losses and associated costs. Real-time insights also allow companies to shift consumption away from peak tariff periods or optimize cooling and heating processes to lower energy bills. The return on investment for advanced monitoring technologies can be rapid, particularly in water-stressed regions where costs and risks are highest. Moreover, the ability to quantify savings strengthens the business case for further efficiency initiatives.

Smart metering also enables greater collaboration across industrial clusters or supply chains. In industrial parks where multiple companies share water infrastructure, advanced monitoring allows for coordinated use, ensuring that shared resources are allocated efficiently and equitably. Supply chain monitoring can extend beyond a company's direct operations, allowing firms to track the water performance of suppliers and work collectively on efficiency improvements. Such collaborative approaches build resilience not just at the company level but across entire networks of production.

Challenges remain in implementing these systems, particularly in terms of upfront costs, integration with legacy infrastructure, and data management. Smaller firms may struggle with the capital investment required, while larger ones may face difficulties integrating new technologies into complex existing systems. Data security and privacy are also critical concerns, as industrial water use data may reveal sensitive information about production processes. Overcoming these challenges requires careful planning, phased implementation, and partnerships with technology providers that can tailor solutions to specific industrial contexts.

Despite these hurdles, the trend toward digitalization in water management is accelerating. As technologies become more affordable and scalable, advanced monitoring and smart metering are likely to become standard practice across industries. Their ability to provide real-time, actionable insights transforms water from a hidden cost into a managed resource, strengthening both operational performance and sustainability outcomes. By embracing these tools, industries can achieve higher levels of efficiency, reduce risks, and demonstrate leadership in addressing global water challenges.

Water-Efficient Equipment and Processes

Water-efficient equipment and processes form the backbone of industrial strategies aimed at reducing consumption while maintaining productivity and product quality. As industries face rising costs, regulatory pressures, and stakeholder scrutiny, investing in equipment designed to minimize water use has become not only environmentally responsible but also economically advantageous. These innovations span production machinery, cleaning and cooling systems, and process redesigns that together enable industries to achieve more output with less water.

One of the most impactful areas of improvement lies in cooling systems. Traditional once-through cooling, where water passes through equipment and is discharged, is highly wasteful. Replacing these systems with closed-loop cooling towers drastically reduces withdrawals by recirculating water. Additional efficiency gains come from hybrid cooling technologies that combine air and water cooling, reducing overall reliance on water without compromising thermal performance. Industries such as power generation, petrochemicals, and metallurgy have already demonstrated significant reductions in water use by adopting these advanced cooling solutions.

Cleaning and sanitation processes represent another major target for water-efficient equipment. In food processing, beverage manufacturing, and pharmaceuticals, high standards of hygiene require frequent cleaning, often resulting in excessive water use. Modern clean-in-place (CIP) systems address this by automating cleaning cycles, optimizing water volumes, and reusing rinse water where feasible. These systems can reduce water use by up to half compared to manual or outdated cleaning methods. High-pressure, low-flow nozzles and dry-cleaning technologies also reduce consumption while maintaining or improving cleaning effectiveness.

In production processes, innovations in equipment design and operation have led to notable efficiency gains. For example, in the textile industry, low-liquor dyeing machines reduce the ratio of water to fabric, cutting water use significantly compared to conventional dyeing methods. In pulp and paper manufacturing, efficient dewatering presses and closed water circuits reduce both water consumption and wastewater volumes. The adoption of membrane filtration, ultrafiltration, and reverse osmosis technologies across various industries has further enabled recycling within processes, ensuring that less fresh water is withdrawn from external sources.

Process optimization complements equipment upgrades. By redesigning workflows, industries can minimize unnecessary steps that consume water. For instance, modifying cooling sequences, adjusting rinse cycles, or adopting counter-current washing—where cleaner water is used in the final rinse stage and progressively reused upstream—can yield significant reductions. Process integration strategies, such as combining heat and water recovery systems, further enhance efficiency by ensuring that energy and water are reused rather than wasted.

Automation and digital control systems enhance the performance of water-efficient equipment. Smart controllers can adjust flow rates, pressure, and temperature in real time, ensuring that equipment uses only as much water as required. Sensors integrated into processes detect variations in demand and quality, reducing overuse and waste. When combined with predictive analytics, these technologies allow companies to anticipate peaks in demand and adjust equipment operation accordingly, aligning efficiency with production schedules.

The benefits of water-efficient equipment and processes extend beyond reduced consumption. They often improve overall operational efficiency, lower energy use, and reduce wastewater treatment costs. For example, when less water enters a system, less wastewater requires treatment, reducing chemical use and disposal fees. These savings can offset initial capital expenditures and deliver strong returns on investment. Additionally, efficient equipment often improves product quality and consistency, strengthening competitiveness.

Barriers remain, particularly for smaller enterprises where the upfront costs of new equipment can be prohibitive. Retrofitting legacy systems may also be technically challenging. However, many governments and development agencies offer financial incentives, subsidies, or technical assistance programs to encourage adoption. As water stress intensifies and technologies become more affordable, the economic case for upgrading equipment becomes even stronger.

Water-efficient equipment and processes, therefore, represent a practical and scalable pathway for industries to reduce their water footprint. By modernizing cooling, cleaning, and production systems, optimizing workflows, and integrating automation, industries can achieve substantial reductions in consumption while realizing financial and operational benefits. These solutions demonstrate that efficiency is not just about conservation but about building resilience, competitiveness, and sustainability into industrial systems.

Treatment and Recycling Technologies

Treatment and recycling technologies are central to industrial efforts to conserve water and reduce dependence on freshwater withdrawals. As water scarcity intensifies and regulations on wastewater discharge become more stringent, industries are investing in advanced solutions that not only ensure compliance but also create opportunities to reuse water within processes. These technologies shift water management from a linear model of "use and dispose" to a circular approach where water is continually treated, recovered, and recycled.

Conventional treatment methods remain foundational. Primary treatment removes solids and debris through screening and sedimentation, while secondary treatment uses biological processes to break down organic matter. For many industries, these steps are essential for meeting baseline discharge requirements. However, conventional systems are often insufficient for enabling reuse, as they may leave behind dissolved salts, trace organics, or other contaminants. To move beyond compliance toward recycling, industries increasingly adopt advanced treatment technologies.

Membrane filtration is one of the most widely used advanced methods. Technologies such as ultrafiltration, nanofiltration, and reverse osmosis provide high levels of purification, removing suspended solids, pathogens, and dissolved salts. Reverse osmosis, in particular, is crucial for industries requiring ultrapure water, such as electronics manufacturing or pharmaceuticals. By integrating these membranes into recycling systems, companies can produce water that is suitable for reuse in cooling, rinsing, or even as process water in highly sensitive applications.

Thermal treatment technologies also play an important role. Evaporation and distillation systems can recover clean water from highly contaminated or saline streams. In some industries, zero-liquid discharge (ZLD) systems combine evaporation with crystallization to eliminate wastewater altogether, recovering both water and solid byproducts that may be reused or safely disposed of. While ZLD can be energy-intensive, advances in system design and the integration of renewable energy are making these technologies more cost-effective and sustainable.

Biological treatment continues to evolve as well. Moving bed biofilm reactors, membrane bioreactors, and aerobic granular sludge systems increase treatment efficiency while reducing footprint and energy use. These systems allow industries to treat wastewater to higher standards, often producing effluent that is suitable for reuse in non-potable applications. For industries with organic-rich wastewater streams, anaerobic digestion not only treats water but also generates biogas, creating an additional energy benefit.

Advanced oxidation processes (AOPs) provide solutions for removing trace contaminants that traditional systems cannot handle. Using combinations of ozone, hydrogen peroxide, or UV light, AOPs break down persistent organic pollutants and micropollutants such as pharmaceuticals or pesticides. This makes them especially valuable in industries where discharge standards are strict and water must be polished to very high levels of purity before reuse.

Industrial recycling systems often integrate multiple treatment technologies in a staged process. For example, a facility may use biological treatment to reduce organic load, membrane filtration to remove solids, and reverse osmosis to produce high-quality water for reuse. The choice of technologies depends on the quality of the incoming wastewater, the intended reuse application, and the economic and environmental context. Increasingly, modular systems are being developed that allow industries to scale and adapt treatment to changing needs.

Digitalization further enhances treatment and recycling. Real-time monitoring systems track water quality parameters, allowing operators to adjust processes instantly and maintain efficiency. Predictive analytics can forecast when membranes require cleaning or when systems are approaching failure, reducing downtime and maintenance costs. Automation ensures consistent performance, particularly in facilities where water quality requirements are strict.

The adoption of treatment and recycling technologies delivers multiple benefits. By reducing dependence on freshwater, industries lower their exposure to scarcity risks and build resilience against droughts and supply disruptions. Recycling also decreases wastewater discharge, reducing the risk of environmental harm and lowering costs associated with disposal and compliance. In many cases, these systems generate long-term financial savings by cutting water purchases and waste management expenses.

While upfront costs and technical complexity can be barriers, the trend toward stricter regulations, higher water tariffs, and sustainability commitments is driving wider adoption. As technologies become more efficient and affordable, they are likely to become standard practice across industries. Treatment and recycling technologies thus represent a cornerstone of industrial water efficiency, enabling companies to meet operational needs while contributing to broader environmental sustainability goals.

Innovations in Cooling and Energy Systems

Cooling and energy systems are among the most water-intensive components of industrial operations, and innovations in this area are transforming how industries manage water use. Traditional cooling approaches, such as once-through systems, rely on large volumes of freshwater that are withdrawn, used briefly, and then discharged. These methods not only strain local water resources but also expose industries to risks during periods of scarcity. Advances in technology and system design are helping companies reduce dependence on freshwater while improving overall energy efficiency, making cooling and energy systems a focal point for industrial water conservation.

Closed-loop cooling systems represent a major improvement over once-through designs. By recirculating water through cooling towers, industries can dramatically cut withdrawals and reuse the same water multiple times. Further innovations in closed-loop systems include hybrid cooling, which combines air and water cooling to reduce overall water demand. This hybrid approach is particularly valuable in regions facing seasonal or chronic water stress, as it allows facilities to switch between water and air cooling depending on availability and cost.

Dry cooling is another significant innovation that nearly eliminates water use by relying on air to dissipate heat. While dry cooling can require higher capital investment and may reduce efficiency during very hot weather, it provides an important option in water-scarce regions where traditional wet cooling is not feasible. Advances in design and materials are making dry cooling more competitive by improving heat transfer efficiency and reducing energy penalties, making it increasingly attractive for sectors such as power generation.

Energy recovery and integration strategies also play an important role in reducing water demand. By capturing and reusing waste heat from industrial processes, companies can reduce the need for additional cooling and the water that comes with it. Combined heat and power (CHP) systems, for example, recover heat that would otherwise be wasted and use it for on-site heating or electricity generation. This reduces overall cooling requirements and improves energy efficiency, creating a synergy that conserves both water and energy resources.

Innovations in advanced materials and coatings are further enhancing cooling system performance. For instance, surfaces treated with nanomaterials or specialized coatings can improve heat transfer while resisting scale and biofouling, both of which reduce efficiency and increase water demand. These improvements extend the lifespan of equipment, reduce maintenance needs, and optimize water use within cooling cycles.

Digital technologies are also transforming cooling and energy system management. Smart sensors, artificial intelligence (AI), and predictive analytics allow operators to monitor temperature, flow, and water quality in real time, enabling precise control of cooling processes. Predictive models can forecast demand peaks and adjust operations accordingly, ensuring that systems use only as much water as necessary. By integrating cooling systems into digital management platforms, industries can continuously optimize water and energy use while quickly identifying and addressing inefficiencies.

Integration with renewable energy is another frontier of innovation. Solar-powered cooling systems, for example, can use renewable energy to power air-based cooling technologies, reducing reliance on water-intensive thermal energy sources. Similarly, coupling renewable energy with advanced treatment and recycling systems can create closed-loop industrial operations that minimize water withdrawals while maintaining energy security.

The benefits of these innovations extend beyond water savings. By reducing cooling water use, industries also decrease the volume of wastewater discharged, cutting costs and environmental risks associated with effluent management. Energy efficiency gains lower greenhouse gas emissions, aligning industrial operations with climate goals and strengthening their ESG performance. Furthermore, adopting cutting-edge cooling and energy systems enhances resilience by reducing exposure to risks associated with water scarcity and rising energy costs.

Challenges to adoption remain, particularly the upfront costs of transitioning to advanced systems and the technical complexities of retrofitting existing facilities. However, growing regulatory pressure, stakeholder expectations, and the clear financial benefits of long-term savings are driving industries to invest in these solutions. As technologies become more scalable and affordable, innovations in cooling and energy systems are poised to become standard practice across industrial sectors.

By rethinking the design and operation of cooling and energy systems, industries can significantly reduce water consumption, lower operational risks, and strengthen resilience in a changing climate. These innovations demonstrate how technical progress can address both environmental and economic challenges, ensuring that industries remain competitive while contributing to global sustainability.

Chapter 5: Water Reuse and Recycling Strategies

Water reuse and recycling strategies are critical components of industrial water management, enabling companies to reduce dependence on freshwater sources while lowering costs and environmental impacts. As freshwater scarcity intensifies and regulatory pressures increase, industries are recognizing that conventional once-through water use is unsustainable. By adopting reuse and recycling practices, companies not only secure operational resilience but also contribute to broader sustainability goals. This chapter examines the approaches, technologies, and management strategies that industries can deploy to maximize the value of every drop, moving from linear to circular models of water use and building long-term resilience.

Closed-Loop and Zero-Liquid Discharge Systems

Closed-loop and zero-liquid discharge (ZLD) systems represent some of the most advanced strategies for reducing industrial water use and minimizing wastewater discharges. Both approaches are central to the transition from linear water management—where water is used once and discharged—to circular systems that maximize reuse, conserve resources, and protect ecosystems. While closed-loop systems focus on recirculating water within processes, ZLD technologies push this approach further by ensuring that no liquid effluent leaves the facility, recovering water and solid byproducts in the process.

Closed-loop systems are designed to keep water in circulation rather than allowing it to be discharged after a single use. In these systems, water is captured after being used for cooling, washing, or production processes, treated as necessary, and then reused in the same or different applications. Cooling towers are one of the most common examples, where water circulates repeatedly, losing only small amounts to evaporation. By recapturing and treating blowdown water, industries can extend the cycle further, reducing withdrawals from external sources. This approach not only conserves freshwater but also cuts costs associated with purchasing, pumping, and treating water.

The benefits of closed-loop systems extend beyond conservation. Recirculating water reduces the volume of effluent requiring treatment, lowering both operational and compliance costs. Because the system is contained, industries also gain greater control over water quality, ensuring that processes relying on specific parameters—such as temperature, pH, or salinity—are not disrupted by fluctuations in external supply. Closed-loop systems are particularly valuable in water-stressed regions where securing reliable withdrawals from rivers or aquifers is difficult or unsustainable.

Zero-liquid discharge systems take circularity one step further. In ZLD, all wastewater generated by industrial processes is treated, purified, and recovered, leaving no liquid waste to be discharged into the environment. Typically, ZLD systems use a combination of treatment technologies such as ultrafiltration, reverse osmosis, evaporation, and crystallization. These processes separate contaminants, recover reusable water, and convert remaining solids into byproducts that may be reused or safely disposed of. For example, salts recovered from crystallizers can sometimes be used in other industrial applications, creating an additional resource stream.

The appeal of ZLD lies in its ability to completely eliminate liquid discharges, protecting water bodies from pollution and ensuring compliance with even the strictest regulatory regimes. In regions where discharge permits are difficult to obtain or water quality standards are particularly stringent, ZLD offers a viable path for industries to maintain operations without risking violations. It also provides resilience by recovering large volumes of water that can be reused internally, reducing dependence on external supplies and insulating industries from droughts or restrictions.

However, ZLD systems are resource- and energy-intensive, often requiring significant capital investment and operating costs. Evaporation and crystallization, in particular, demand large amounts of energy, which can increase the carbon footprint of a facility if not managed carefully. Innovations are addressing these challenges, with new designs incorporating more efficient membranes, multi-effect evaporators, and integration with renewable energy to reduce overall costs and emissions. Modular ZLD systems are also being developed, allowing industries to scale implementation based on their size, wastewater characteristics, and budget.

The decision between implementing a closed-loop system or a full ZLD system often depends on the local regulatory environment, the quality of incoming and outgoing water, and the economic trade-offs. In some cases, hybrid systems are adopted, where water is largely recirculated through closed-loop systems, and remaining effluent streams are treated with ZLD technologies to achieve near-total reuse. These integrated approaches balance cost-effectiveness with environmental performance, allowing industries to significantly reduce both withdrawals and discharges.

Digital monitoring and automation are playing a growing role in the success of closed-loop and ZLD systems. Smart sensors track flow rates, pressure, and quality parameters in real time, ensuring that treatment systems operate at peak efficiency. Predictive analytics help anticipate fouling or scaling in membranes and evaporators, allowing for timely maintenance and reduced downtime. Automation ensures consistent performance, which is particularly important in ZLD systems where multiple treatment steps must be coordinated seamlessly to avoid bottlenecks.

The broader benefits of closed-loop and ZLD systems include improved corporate reputation and stronger stakeholder relationships. By demonstrating leadership in water stewardship, industries can gain community trust, meet investor expectations, and strengthen compliance with ESG reporting requirements. These systems also contribute to broader sustainability goals by protecting ecosystems, reducing pollution, and aligning with global agendas such as the UN Sustainable Development Goals.

Closed-loop and ZLD systems illustrate how industries can transform water management from a linear model of consumption and disposal into a circular framework of reuse and recovery. While cost and energy demands remain challenges, continued innovation and regulatory pressures are driving wider adoption. As water scarcity and quality concerns grow more urgent, these systems are likely to become increasingly standard in industrial practice, representing a key pathway toward sustainable and resilient operations.

Industrial Symbiosis and Byproduct Water Reuse

Industrial symbiosis and byproduct water reuse represent innovative approaches that extend water efficiency beyond individual facilities by promoting collaboration across industries. Instead of viewing wastewater as a burden, these strategies recognize it as a resource that can be shared, treated, and repurposed to support broader industrial ecosystems. This systems-level thinking transforms industrial parks and clusters into networks where water, energy, and material flows are optimized collectively, reducing overall demand on freshwater supplies and lowering environmental impacts.

Industrial symbiosis is based on the principle that the waste output of one facility can become the input for another. In the context of water, this often means that the byproduct water from one process, after appropriate treatment, is reused by a neighboring facility for cooling, washing, or even as process water. For example, a power plant might provide warm wastewater to an adjacent agricultural or aquaculture enterprise, or a manufacturing facility may supply treated effluent to another industry that requires non-potable water. These exchanges reduce the need for each facility to secure its own freshwater supply, creating efficiency gains across the cluster.

Byproduct water reuse also plays a significant role in minimizing discharges to the environment. Instead of releasing treated effluent into rivers or aquifers, industries can retain this water within the industrial system, thereby lowering pollution risks while maximizing the utility of every unit withdrawn. This approach not only conserves scarce resources but also reduces regulatory and community pressures associated with wastewater discharges. In some cases, byproduct reuse can extend beyond water to include recovery of salts, nutrients, or chemicals that can be repurposed, further enhancing the circularity of industrial processes.

The success of industrial symbiosis depends heavily on infrastructure and governance. Shared pipelines, storage facilities, and treatment plants are often necessary to enable water exchanges between companies. Governance structures, such as industrial park authorities or public-private partnerships, play a key role in coordinating these flows, setting quality standards, and ensuring equitable allocation of costs and benefits. Without such coordination, logistical challenges and concerns over liability or water quality may hinder collaboration.

Technological advances are making industrial symbiosis more feasible. Modular treatment units, advanced filtration systems, and smart monitoring technologies allow water to be tailored to the specific needs of different users. For example, one facility may require high-purity water for production, while another can make use of lower-quality effluent for cooling. Advanced treatment ensures that byproduct water can be matched to the right application, expanding the range of potential symbiotic exchanges.

Economic incentives further drive adoption. Sharing water resources can reduce costs for all participants by lowering withdrawals, treatment, and discharge expenses. Companies also benefit from enhanced resilience, as symbiotic networks diversify supply sources and reduce vulnerability to scarcity. In some cases, industrial parks have marketed themselves as sustainability leaders by highlighting their water reuse systems, attracting tenants and investors who prioritize environmental performance.

Community and regulatory pressures also support the growth of industrial symbiosis. Communities are more likely to support industrial development when companies demonstrate collaborative and responsible water use. Regulators may encourage symbiosis by offering incentives, streamlined permitting, or recognition programs for shared water systems. These external drivers align with internal economic and environmental benefits, reinforcing the case for adoption.

While the potential is significant, challenges remain. Coordinating multiple stakeholders with different priorities, ensuring consistent water quality, and managing costs equitably require strong collaboration and trust. Legal frameworks must clarify ownership, liability, and compliance responsibilities for shared water resources. Overcoming these challenges requires transparent agreements, supportive policy environments, and effective governance structures that balance the interests of all parties.

Industrial symbiosis and byproduct water reuse offer industries a pathway to achieve higher levels of water efficiency while building resilience and sustainability at the system level. By treating wastewater as a resource and fostering collaboration across sectors, these approaches reduce pressure on freshwater ecosystems, lower costs, and strengthen social license to operate. As water scarcity and climate variability intensify, such cooperative and circular models are likely to become increasingly important for industries worldwide.

Wastewater Treatment for Reuse

Wastewater treatment for reuse has become a central pillar of industrial water management strategies, enabling companies to reduce their reliance on freshwater withdrawals while minimizing pollution discharges. Rather than viewing wastewater as an end product to be disposed of, industries are increasingly treating it as a valuable resource that can be purified and cycled back into operations. This approach supports both operational efficiency and sustainability, making it a key pathway for industries operating in water-stressed environments or under stringent regulations.

The treatment process for reuse typically involves multiple stages, each designed to remove specific types of contaminants and prepare the water for its intended application. Primary treatment removes larger solids and sediments through screening and settling. Secondary treatment uses biological processes, such as activated sludge systems or biofilm reactors, to degrade organic matter and reduce biochemical oxygen demand. For many non-potable applications, these first two stages may be sufficient, especially when the treated water is destined for cooling, washing, or irrigation.

However, higher-value reuse often requires advanced treatment. Tertiary processes such as membrane filtration, reverse osmosis, and advanced oxidation are used to remove dissolved salts, nutrients, trace organics, and pathogens. Industries with strict quality requirements—such as pharmaceuticals, electronics, and food processing—depend on these technologies to ensure that recycled water meets rigorous standards. Membrane bioreactors (MBRs) combine biological treatment with ultrafiltration, producing high-quality effluent in a compact footprint, making them increasingly popular in industrial applications.

The reuse of treated wastewater takes many forms depending on the industry and local conditions. Cooling and boiler feedwater are common applications, as both require large volumes of water that do not necessarily need to be potable. In manufacturing, recycled water can be used for washing raw materials, rinsing products, or cleaning equipment. In some cases, treated wastewater can even be reintroduced directly into production processes when advanced treatment technologies ensure sufficient quality. Beyond the plant boundary, some industries supply surplus treated water to municipalities, agriculture, or neighboring businesses, contributing to broader water security.

The economic case for wastewater reuse is increasingly compelling. While the capital costs of advanced treatment technologies can be high, the savings from reduced freshwater purchases, lower discharge fees, and decreased energy use often deliver attractive payback periods. Furthermore, reuse reduces exposure to risks associated with droughts, restrictions, or competition for shared water resources. In regions where water tariffs are rising or regulations are tightening, investments in treatment for reuse are becoming less an option and more a necessity.

Environmental benefits are equally significant. Recycling wastewater reduces effluent discharges into rivers, lakes, and aquifers, lowering the burden of pollution on ecosystems and communities. It also lessens pressure on natural freshwater sources, helping to preserve biodiversity and ensure that water remains available for households and agriculture. By reducing both withdrawals and discharges, wastewater reuse embodies the principles of the circular economy, turning waste into a resource and closing the loop on industrial water use.

The adoption of wastewater treatment for reuse is being accelerated by digital technologies. Real-time monitoring and control systems ensure that treatment plants operate efficiently and consistently, producing water of reliable quality. Predictive maintenance powered by AI helps operators anticipate equipment failures and optimize performance. These digital tools enhance confidence in reuse systems, especially in industries where even small variations in water quality can disrupt production or compromise product safety.

Despite these advantages, challenges remain. Smaller enterprises may struggle with the capital costs of advanced treatment infrastructure, and industries operating in regions with weak regulatory frameworks may lack incentives to invest. Concerns over liability, especially when treated water is shared across industries or supplied to external users, also require clear governance and robust quality assurance systems. Addressing these challenges calls for supportive policies, financing mechanisms, and public-private partnerships that lower barriers to adoption.

Wastewater treatment for reuse is more than a technical solution; it represents a shift in mindset from linear to circular resource use. By embracing this approach, industries not only reduce operational risks and costs but also contribute to broader goals of environmental protection and sustainable development. As pressures on global water resources intensify, wastewater reuse is set to become a defining feature of industrial water strategies, ensuring that water is valued, conserved, and used to its fullest potential.

Integration with Circular Economy Approaches

Integrating water conservation and efficiency with circular economy approaches is transforming how industries view and manage water. Traditionally, water has been treated as a one-way input: it is withdrawn, used in production, and discharged as wastewater. The circular economy challenges this linear model by emphasizing resource recovery, reuse, and closed-loop systems that minimize waste and maximize value. For industries, adopting circular economy principles means treating water not as a disposable commodity but as a resource to be conserved, regenerated, and cycled continuously through operations.

A central aspect of circular water management is designing systems where wastewater becomes a resource rather than a liability. By investing in advanced treatment technologies, industries can recover water of sufficient quality for reuse within the same facility or across different operations. For example, rinse water from one production line can be treated and repurposed for cooling in another, reducing both freshwater withdrawals and effluent discharges. Beyond water itself, circular approaches encourage the recovery of valuable byproducts—such as nutrients, salts, or energy—from wastewater streams, turning what was once waste into inputs for other processes.

Industrial parks and eco-industrial clusters exemplify how circular economy principles can be applied at scale. In these settings, multiple industries collaborate to share resources, including water. The effluent from one company can serve as an input for another, creating symbiotic exchanges that reduce overall demand on external water supplies. For instance, treated wastewater from a chemical plant could be used by a nearby power facility for cooling, while recovered heat or energy from industrial processes supports other operations. Such integration fosters systemic efficiency that no single company could achieve in isolation.

The circular economy also emphasizes designing products and processes that reduce water demand throughout their life cycles. For industries, this means not only focusing on internal operations but also considering how water is embedded in supply chains and product use. By collaborating with suppliers and customers, companies can reduce water footprints across the entire value chain. Examples include encouraging suppliers to adopt water-efficient practices or designing products that require less water for consumer use and disposal. These efforts extend the principles of circularity beyond factory walls, embedding them into broader industrial ecosystems.

Digitalization strengthens the integration of circular economy approaches with water management. Smart sensors, IoT platforms, and data analytics provide real-time insights into water flows, making it easier to identify opportunities for reuse and recovery. Predictive models can optimize when and how water should be cycled through different processes, while blockchain and traceability systems ensure accountability across supply chains. This technological layer supports transparency and builds trust among stakeholders, enabling more ambitious circular strategies to be implemented collaboratively.

Economic and environmental benefits reinforce the appeal of circular water management. By reducing reliance on freshwater withdrawals, industries lower operational risks associated with scarcity and regulatory restrictions. Reusing water decreases wastewater discharge volumes, cutting treatment and compliance costs. Recovering byproducts creates new revenue streams or offsets material purchases. Together, these advantages create a strong business case for integrating circular economy principles, especially in regions where water is scarce, costs are high, or regulations are tightening.

Barriers to integration remain, including the high upfront costs of advanced treatment systems, limited infrastructure for resource exchange, and institutional resistance to collaboration between companies. Overcoming these obstacles requires supportive policy frameworks, incentives for innovation, and governance structures that encourage cooperation. Governments and industry associations can play a pivotal role in creating the conditions for circular models to thrive, including setting reuse targets, offering tax incentives, or facilitating partnerships.

Integration with circular economy approaches redefines water management as part of a broader effort to redesign industrial systems for sustainability. By embedding reuse, recovery, and collaboration into operations, industries can significantly reduce their water footprints while enhancing resilience and competitiveness. This shift reflects a deeper cultural change: from seeing water as a consumable input to valuing it as a renewable resource that underpins both industrial productivity and environmental health. As global pressures on water resources intensify, industries that adopt circular principles will be better prepared to meet the challenges of a resource-constrained future.

Chapter 6: Governance and Management Approaches

Effective governance and management approaches are essential for embedding water conservation and efficiency into the core of industrial strategy. Industries face increasing scrutiny from regulators, investors, and communities, requiring more structured and transparent systems of oversight. Governance provides the framework for accountability, ensuring that responsibilities are clearly defined, while management practices translate strategic commitments into operational action. This chapter explores how companies can build robust governance structures, integrate water into enterprise risk management, and link performance to ESG reporting. Together, these approaches create a culture of responsibility and resilience, positioning industries to meet both current and future water challenges.

Corporate Water Governance Structures

Corporate water governance structures are the frameworks, policies, and organizational arrangements that determine how companies manage water as a strategic resource. In an era of intensifying water scarcity, climate variability, and growing stakeholder expectations, governance provides the foundation for aligning business operations with sustainable water use. Without clear structures in place, companies risk fragmented approaches that fail to address water-related risks or capitalize on opportunities for efficiency, resilience, and reputation building. Effective governance integrates water into corporate strategy, defines accountability, and ensures that decision-making reflects both business needs and societal responsibilities.

At the core of governance is board-level oversight. Increasingly, corporate boards are being called upon to treat water as a material issue that affects financial performance and long-term viability. Board committees on sustainability, risk, or ESG (environmental, social, and governance) often oversee water-related policies, ensuring that they are integrated into enterprise risk management frameworks. By embedding water in board discussions, companies signal to investors and stakeholders that it is a priority issue, on par with climate, energy, and financial risks. This top-level accountability sets the tone for the entire organization.

Clear policies and commitments form the next layer of governance. Many companies establish corporate water policies that articulate principles such as efficiency, reuse, quality protection, and stewardship. These policies may commit the organization to measurable targets—for example, reducing water intensity by a certain percentage, achieving zero liquid discharge at key facilities, or ensuring compliance with global standards. Policies provide consistency across global operations, ensuring that facilities in different regions adhere to common expectations while allowing for adaptation to local conditions.

Organizational structures are essential for translating policies into action. Companies often create cross-functional water management teams that include representatives from operations, sustainability, finance, and supply chain departments. These teams ensure that water-related decisions are not siloed within environmental units but are integrated into broader business processes. Dedicated water managers or sustainability officers may coordinate initiatives, track performance, and liaise with external stakeholders. In large multinational firms, regional or facility-level coordinators play a critical role in adapting global policies to local realities.

Governance also requires robust measurement and reporting systems. Companies must establish protocols for monitoring water withdrawals, use, recycling, and discharge across their operations. Standardized metrics enable consistent tracking and comparison, while digital platforms provide real-time insights. Reporting frameworks such as the GRI, CDP Water Security, and the Sustainability Accounting Standards Board (SASB) guide companies in disclosing performance to investors and stakeholders. Transparent reporting reinforces accountability, while also helping companies benchmark against peers and identify areas for improvement.

Risk assessment is another key element of governance. Companies need systems to evaluate how water scarcity, flooding, pollution, or regulatory changes could affect operations, supply chains, and markets. Integrating water risk into enterprise risk management ensures that decisions on investments, site selection, or production planning account for water-related vulnerabilities. For example, companies may avoid siting new facilities in highly water-stressed regions or prioritize efficiency investments at facilities facing the greatest risks. Strong governance structures ensure that such assessments are systematic, data-driven, and linked to decision-making at the highest levels.

External engagement is increasingly recognized as part of corporate water governance. Companies operate within shared watersheds and must collaborate with governments, communities, and other users to ensure long-term water security. Governance frameworks should define how the company engages in collective action, advocates for sound water policy, and addresses community concerns. Participation in initiatives such as the CEO Water Mandate or the Alliance for Water Stewardship demonstrates leadership and provides companies with guidance on best practices. Governance structures that emphasize external engagement help secure a social license to operate and reduce conflict over shared resources.

Incentives and accountability mechanisms strengthen governance by aligning employee behavior with corporate water goals. Performance metrics tied to water use can be integrated into management evaluations, bonuses, or operational scorecards. Training programs ensure that employees at all levels understand the importance of water efficiency and their role in achieving corporate targets. By embedding accountability throughout the organization, companies ensure that water governance is not limited to high-level policy but is reflected in day-to-day practices.

Challenges remain in designing and implementing effective water governance structures. Many companies operate across multiple jurisdictions with varying regulations and water risks, making it difficult to apply a uniform approach. Balancing global consistency with local flexibility requires careful governance design. Additionally, integrating water governance with other sustainability priorities, such as carbon reduction or biodiversity protection, demands cross-functional coordination that not all organizations are prepared to manage.

Despite these challenges, corporate water governance structures are becoming increasingly sophisticated. Companies that establish clear oversight, strong policies, cross-functional teams, robust monitoring, and external engagement are better positioned to manage risks, capture opportunities, and demonstrate leadership in sustainability. As water scarcity and climate pressures intensify, governance will remain the foundation on which industries build resilience, efficiency, and trust. By embedding water governance into corporate strategy, companies ensure that water is not only managed responsibly but also valued as a resource critical to long-term success.

Risk Assessment and Water Accounting

Risk assessment and water accounting are essential tools for industries to understand, manage, and mitigate the challenges associated with water use. As water scarcity intensifies and regulatory as well as stakeholder expectations grow, companies must take a systematic approach to identifying risks and quantifying their water footprint. These tools enable organizations to anticipate vulnerabilities, plan effective interventions, and demonstrate accountability to investors, regulators, and communities.

Risk assessment begins with understanding the context in which a facility operates. Industries must evaluate both physical and non-physical risks. Physical risks include water scarcity, droughts, flooding, and declining water quality, all of which can directly disrupt operations. Non-physical risks include regulatory changes, reputational impacts, and financial risks associated with tariffs, penalties, or supply chain disruptions. A robust risk assessment maps out these threats across local, regional, and global levels, ensuring that companies account for the full spectrum of challenges they may encounter.

A critical part of this process is assessing basin-level risks. Companies need to evaluate the availability and quality of water resources in the watersheds where they operate. Tools such as the World Resources Institute's Aqueduct or the WWF Water Risk Filter help businesses identify regions with high stress levels, flooding potential, or governance weaknesses. Understanding basin conditions allows companies to anticipate constraints and design strategies that align with local realities, reducing the likelihood of conflicts with other water users.

Water accounting complements risk assessment by quantifying how much water a company uses, withdraws, and discharges. It provides a transparent picture of water flows within operations, helping industries identify inefficiencies and opportunities for conservation. Accounting frameworks often distinguish between different types of water use: blue water (surface and groundwater withdrawals), green water (rainwater embedded in supply chains, particularly agriculture), and grey water (the volume of freshwater required to assimilate pollutants). By breaking water use into these categories, companies can capture the full scope of their footprint across operations and value chains.

Standardized approaches to water accounting are critical for consistency and comparability. Protocols such as the ISO 14046 Water Footprint Standard or the Global Water Accounting Framework offer methodologies for measuring and reporting water use. These frameworks ensure that companies can evaluate their performance over time, benchmark against peers, and disclose results in line with investor expectations. Water accounting also supports decision-making by highlighting hotspots where reductions can yield the greatest environmental and financial benefits.

Integrating risk assessment with water accounting provides industries with actionable insights. For example, a facility may identify that its largest water withdrawals occur in a high-risk basin with increasing drought frequency. Combining these insights highlights both the operational vulnerability and the opportunity to invest in efficiency or alternative sourcing. Similarly, accounting data on wastewater discharges can be paired with risk analysis to assess reputational or regulatory exposure, guiding investments in treatment and recycling technologies.

Financial and strategic planning increasingly rely on this integration. Investors and lenders evaluate corporate resilience by examining both water risk exposure and the robustness of accounting practices. Companies that fail to assess risks or disclose water use transparently may face higher borrowing costs or reduced access to capital. Conversely, those that demonstrate comprehensive risk management and accounting can strengthen investor confidence and gain competitive advantages.

Internal governance structures benefit as well. Risk assessments inform enterprise risk management frameworks, ensuring that water-related vulnerabilities are considered alongside other business risks such as supply chain disruptions or energy costs. Water accounting provides the metrics needed to set and track performance against corporate water targets, embedding efficiency into strategic planning and operational decision-making.

Challenges remain in implementing these practices consistently. Collecting accurate data across complex operations and supply chains can be difficult, particularly in regions with limited monitoring infrastructure. Companies may also face uncertainty in predicting climate impacts or translating basin-level risks into actionable business strategies. Nonetheless, advances in digital technologies, such as remote sensing, IoT sensors, and data analytics, are improving the accuracy and accessibility of both risk assessment and accounting.

Together, risk assessment and water accounting form the backbone of industrial water management. They provide the information needed to anticipate challenges, prioritize investments, and demonstrate accountability to stakeholders. By systematically identifying risks and quantifying water use, industries can reduce vulnerabilities, improve efficiency, and position themselves as leaders in sustainable resource management. As pressures on global water resources continue to mount, these tools will remain indispensable for building industrial resilience in a rapidly changing world.

Integrating Water into Enterprise Risk Management

Integrating water into enterprise risk management (ERM) ensures that companies treat water as a strategic issue rather than a narrow operational concern. Traditionally, ERM frameworks have focused on financial, operational, and compliance risks, but growing recognition of environmental and social challenges has expanded their scope. Water scarcity, flooding, pollution, and regulatory uncertainty are now acknowledged as material risks that can disrupt production, increase costs, and damage reputations. By embedding water considerations into ERM, industries can proactively manage vulnerabilities and strengthen long-term resilience.

The first step is recognizing water as a cross-cutting risk that affects multiple aspects of business performance. Physical risks, such as droughts or declining groundwater levels, directly threaten the continuity of operations. Regulatory risks emerge when governments tighten standards for withdrawals, effluent discharge, or efficiency. Reputational risks arise when companies are perceived as irresponsible water users, potentially triggering consumer backlash or community opposition. Financial risks appear in the form of rising tariffs, penalties, or stranded assets in water-stressed regions. ERM frameworks allow companies to capture these diverse risk categories in a structured and integrated way.

Embedding water into ERM requires systematic risk identification and assessment. Companies can use tools such as scenario analysis, stress testing, and basin-level risk mapping to understand how water challenges might evolve under different climate or regulatory futures. For example, scenario modeling may reveal that a facility located in a semi-arid region could face severe production disruptions within a decade if droughts intensify. Incorporating such findings into ERM ensures that senior management and boards are aware of potential risks and can take action before they materialize.

Quantification and prioritization are essential for making water risks actionable within ERM. Using metrics such as water intensity, dependency ratios, or potential financial losses, companies can rank risks based on their likelihood and impact. This enables managers to allocate resources efficiently, focusing on the most material threats. For instance, a facility with high water withdrawals in a stressed basin may be prioritized for investments in recycling or efficiency technologies, while a plant in a water-abundant region may require only monitoring and contingency planning.

Once identified and prioritized, water risks must be integrated into enterprise-wide strategies and controls. This includes updating corporate policies, supply chain management practices, and investment decisions. In supply chains, for example, ERM processes can highlight vulnerabilities where key suppliers operate in water-scarce regions, prompting diversification of sourcing or collaborative efficiency initiatives. In capital planning, ERM can guide site selection for new facilities by factoring water availability and regulatory stability into decision criteria. In this way, water becomes a lens through which strategic choices are evaluated.

Monitoring and reporting are critical for maintaining the integration of water into ERM. Companies should establish indicators that track progress in managing water-related risks and include them in regular risk reports to executives and boards. Disclosure frameworks such as CDP Water Security or the Task Force on Climate-related Financial Disclosures (TCFD) provide structured ways to communicate water risks to investors and stakeholders. Transparent reporting strengthens accountability and builds trust, while also aligning with investor expectations that water risks be treated alongside financial and climate risks.

Integrating water into ERM also involves cultivating a culture of risk awareness across the organization. Training programs can help employees at all levels understand the importance of water risks and their role in managing them. Cross-functional teams that bring together sustainability, finance, operations, and supply chain specialists ensure that water is considered in decision-making across departments. When water becomes a shared responsibility, companies are better equipped to anticipate challenges and respond swiftly when conditions change.

Although challenges exist—such as data limitations, uncertainty in climate projections, and the complexity of global supply chains—the benefits of integrating water into ERM are substantial. Companies gain foresight into potential disruptions, reduce financial volatility, and enhance their ability to meet regulatory and stakeholder expectations. By embedding water risks into enterprise frameworks, industries move beyond reactive management to proactive strategy, ensuring that water security is seen not just as an environmental issue but as a core determinant of business resilience and long-term success.

Linking Water Management to ESG Reporting

Linking water management to ESG reporting has become a critical component of corporate sustainability. As investors, regulators, and stakeholders demand greater transparency, water-related risks and performance are increasingly recognized as material factors influencing long-term business resilience. ESG reporting provides a structured platform for companies to disclose how they manage water consumption, quality, and risks, ensuring that these efforts are visible, comparable, and aligned with broader sustainability agendas.

Water is directly connected to the "E" in ESG, where it sits alongside issues like climate change and biodiversity. Companies are expected to demonstrate how they are reducing withdrawals, increasing efficiency, reusing wastewater, and protecting ecosystems. Metrics such as total withdrawals, discharge volumes, water intensity per unit of output, and percentage of water recycled are commonly reported. These quantitative indicators provide a clear picture of performance, but they are most meaningful when combined with narrative disclosures explaining strategies, governance, and long-term commitments. Strong ESG reporting ensures that water management is not only tracked internally but also communicated externally as a core element of environmental stewardship.

The "S" dimension of ESG highlights the social implications of water use. Industries often operate in shared watersheds where competition for limited resources can affect local communities. Transparent reporting on community engagement, stakeholder partnerships, and equitable water allocation helps demonstrate a company's commitment to social responsibility. For example, disclosing efforts to minimize conflicts with local users, ensure safe drinking water access, or invest in watershed restoration strengthens trust and legitimacy. By linking water management to social outcomes, companies show how their actions contribute to both operational stability and societal well-being.

Governance, the "G" in ESG, reinforces accountability for water management. Effective reporting highlights the governance structures that oversee water performance, such as board-level committees, cross-functional management teams, and risk oversight frameworks. Investors and stakeholders want to see that water is treated as a strategic issue, not just a technical concern. ESG disclosures should outline responsibilities, decision-making processes, and integration of water risks into enterprise risk management. This governance transparency assures stakeholders that companies are equipped to anticipate and manage water-related challenges.

Frameworks and standards guide how companies link water to ESG reporting. The GRI provides detailed indicators on water withdrawals, discharges, and impacts. The SASB tailors disclosures to sector-specific materiality, such as water use in manufacturing, agriculture, or extractive industries. CDP Water Security questionnaires offer investors consistent data on risks, strategies, and performance. Emerging frameworks like the Taskforce on Nature-related Financial Disclosures (TNFD) will further shape how companies report on water as part of broader natural capital dependencies. Aligning with these frameworks ensures comparability, credibility, and alignment with global best practices.

Digital tools enhance the integration of water management into ESG reporting. Advanced monitoring systems, smart meters, and data platforms generate reliable, real-time data that can be directly incorporated into ESG disclosures. Predictive analytics allow companies to demonstrate forward-looking risk management, while blockchain and traceability tools can strengthen supply chain transparency. These technologies not only improve internal performance tracking but also lend credibility to external reporting, addressing concerns about greenwashing.

The benefits of linking water management to ESG reporting are both reputational and financial. Transparent disclosures improve access to capital by meeting investor expectations and reducing perceived risk. Many institutional investors now require water-related data to evaluate resilience, and companies with strong reporting often enjoy lower borrowing costs. ESG-linked loans and sustainability bonds also depend on credible disclosures, providing a direct financial incentive. Additionally, public reporting enhances brand value, strengthens stakeholder trust, and reduces the risk of reputational damage from water-related conflicts or crises.

Challenges include inconsistent data quality across operations, lack of standardization among reporting frameworks, and the complexity of accounting for supply chain water use. Overcoming these requires building robust internal data systems, aligning with widely recognized standards, and engaging suppliers in collaborative reporting. Companies must also ensure that reporting is not a compliance exercise but reflects genuine integration of water into strategy and culture.

By linking water management to ESG reporting, companies demonstrate that they view water not only as a resource to be managed but as a strategic factor shaping resilience, competitiveness, and legitimacy. Strong disclosures make water performance visible to stakeholders, align companies with global sustainability agendas, and drive accountability throughout operations and supply chains. As ESG reporting becomes more rigorous and standardized, industries that integrate water effectively into their disclosures will be better positioned to attract capital, build trust, and thrive in a resource-constrained world.

Chapter 7: Cross-Sectoral and Supply Chain Perspectives

Industrial water efficiency cannot be addressed in isolation, as industries are interconnected with agriculture, energy, municipalities, and global supply chains. Water challenges often extend beyond facility boundaries, requiring companies to recognize and act on the broader networks in which they operate. Cross-sectoral collaboration enables industries to share resources, develop joint infrastructure, and align strategies with other water users, reducing competition and fostering resilience. Similarly, supply chains represent both risks and opportunities, as upstream and downstream partners influence overall water footprints. This chapter explores how industries can integrate cross-sectoral perspectives and supply chain considerations into their water strategies, ensuring that conservation and efficiency extend across entire value chains.

Water Efficiency in Supply Chains

Water efficiency in supply chains has become a critical dimension of industrial sustainability, as companies increasingly recognize that their greatest water impacts often lie beyond their direct operations. For many industries, raw material extraction, processing, and transportation account for the bulk of water use, dwarfing the volumes consumed within factories or plants. By addressing water efficiency across the supply chain, businesses not only reduce risks to their own operations but also contribute to global efforts to manage water scarcity and protect ecosystems.

A first step is mapping water use across the supply chain. This requires understanding where suppliers are located, the water demands of their processes, and the water stress levels of those regions. Agricultural supply chains, for instance, often rely on large volumes of irrigation water, making them highly vulnerable to droughts and competing demands. Extractive industries such as mining and chemicals similarly have high water dependencies, often situated in arid or semi-arid areas. By identifying hotspots of water use and stress, companies can prioritize interventions where they will have the greatest impact.

Water accounting frameworks support this mapping exercise. Tools such as life cycle assessment (LCA) and water footprint analysis quantify the embedded water—sometimes referred to as "virtual water"—in raw materials and products. These tools allow companies to compare the water intensity of different suppliers or sourcing options, guiding procurement decisions. For example, a beverage company might assess whether to source sugar from a region with high water availability rather than one experiencing chronic scarcity, thereby reducing both operational risks and reputational concerns.

Collaboration with suppliers is central to improving supply chain water efficiency. Many suppliers, particularly small and medium-sized enterprises (SMEs), may lack the capacity to invest in advanced water technologies. By providing technical assistance, financial incentives, or access to efficiency tools, larger companies can help suppliers adopt water-saving practices. This collaboration creates shared value: suppliers reduce costs and risks, while buyers secure more resilient supply chains. Some industries have established collective programs that bring together multiple companies to support suppliers in high-risk regions, amplifying the impact of interventions.

Certification and standards play a complementary role. Programs such as the Alliance for Water Stewardship (AWS) or agricultural standards like Rainforest Alliance encourage suppliers to adopt best practices in water management. By requiring or incentivizing suppliers to achieve certification, companies can create consistent expectations across their supply chains and provide assurance to investors and consumers that water is being managed responsibly. These standards also encourage continuous improvement, ensuring that suppliers remain aligned with evolving sustainability goals.

Digitalization enhances supply chain water efficiency by enabling greater transparency and accountability. Remote sensing, blockchain, and IoT technologies can track water use in real time, providing data on where and how water is consumed throughout the supply chain. These technologies make it easier to verify supplier performance, detect risks early, and demonstrate compliance with reporting frameworks. Digital platforms also facilitate collaboration by connecting suppliers with buyers and stakeholders in shared water basins.

Financial drivers reinforce the importance of supply chain water efficiency. Investors are increasingly scrutinizing water risks as part of ESG assessments, and companies with highly water-intensive supply chains may face higher borrowing costs or reputational risks. By addressing water use beyond direct operations, companies demonstrate a proactive approach to risk management, which strengthens investor confidence. Efficiency also reduces costs in the long term, as suppliers that use less water often spend less on energy, treatment, and compliance.

Social and community dimensions further emphasize the need for supply chain action. Many suppliers operate in regions where communities already face water scarcity. When industries fail to address water use in these areas, conflicts with local populations can arise, damaging reputations and threatening licenses to operate. Conversely, by promoting water efficiency and stewardship across supply chains, companies contribute to shared water security, strengthening their social license and building positive relationships with communities.

Despite its importance, supply chain water efficiency presents challenges. Data collection can be difficult, especially when supply chains are complex and global. Suppliers may be reluctant to share information or invest in improvements without clear incentives. Overcoming these barriers requires strong governance, transparent communication, and the creation of business models where all parties benefit from efficiency gains.

Water efficiency in supply chains is no longer optional—it is a strategic imperative. By mapping water use, collaborating with suppliers, leveraging standards, and embracing digital tools, industries can significantly reduce their indirect water footprints. These actions not only conserve resources but also build resilience, enhance competitiveness, and contribute to the long-term sustainability of global supply networks. As water challenges intensify, companies that embed efficiency throughout their supply chains will be better positioned to thrive in a resource-constrained future.

Collaboration with Agriculture and Energy Sectors

Collaboration between industry, agriculture, and energy sectors is essential for achieving large-scale water efficiency and resilience. These three sectors collectively account for the vast majority of global freshwater withdrawals, and their demands often intersect within the same river basins and aquifers. When each sector acts independently, competition for scarce water resources can lead to inefficiencies, conflicts, and environmental degradation. By working together, industries can create synergies that conserve water, reduce costs, and support sustainable development at the basin scale.

The agriculture sector is the world's largest water user, responsible for roughly 70 percent of withdrawals, primarily for irrigation. Industries that depend on agricultural commodities—such as food and beverage, textiles, and biofuels—are directly tied to this water demand through their supply chains. Collaborating with agriculture offers opportunities to promote more efficient irrigation practices, crop choices, and soil management techniques that reduce water consumption. For example, industries can support the adoption of drip irrigation, soil moisture monitoring, or drought-resistant crop varieties among their suppliers. Such efforts not only conserve water but also strengthen supply chain resilience, ensuring that raw materials remain available even during times of scarcity.

Partnerships with agriculture also extend to shared infrastructure and wastewater reuse. In some regions, treated industrial effluent can be safely reused for irrigation, reducing pressure on freshwater sources. Conversely, agricultural drainage water can sometimes be treated and recycled for industrial use, creating circular flows between the two sectors. Collaborative investments in shared treatment facilities, reservoirs, or distribution systems can generate cost savings and water security benefits for both farmers and industries. These joint initiatives also strengthen relationships with local communities, demonstrating that industries are contributing positively to rural development.

The energy sector represents another major area of water demand, particularly in thermal power generation and refining. Cooling processes for fossil fuel and nuclear power plants require large volumes of water, while hydropower directly depends on river flows. As industries consume electricity and fuels produced by this sector, their indirect water footprint is significant. Collaboration with the energy sector can focus on reducing the water intensity of power production and aligning industrial demand with less water-intensive sources of energy. For example, industries can partner with utilities to expand renewable energy capacity, such as solar or wind, which require little to no water for operation.

Industrial facilities that produce excess heat or wastewater can also collaborate with the energy sector through cogeneration and resource exchange. Waste heat from factories can support district heating systems, reducing the need for water-intensive power generation. Similarly, biogas generated from anaerobic wastewater treatment can be fed into energy systems, displacing fossil fuel use and reducing overall water and carbon footprints. These symbiotic relationships exemplify how cross-sector collaboration can unlock efficiencies that no sector could achieve alone.

Governance and policy frameworks are vital to enabling these collaborations. Integrated water resources management (IWRM) provides a structure for aligning industrial, agricultural, and energy demands at the basin scale. By participating in watershed councils, multi-stakeholder platforms, or public-private partnerships, industries can engage with farmers, utilities, and governments to balance competing needs and co-design solutions. Policy incentives, such as subsidies for joint efficiency projects or recognition programs for cross-sector initiatives, can further encourage collaboration.

Digital technologies provide tools to support cross-sectoral collaboration. Basin-level monitoring platforms, remote sensing, and shared data systems enable stakeholders to understand cumulative water use and impacts. These insights help identify opportunities for joint efficiency projects, monitor outcomes, and ensure transparency. For example, real-time data on agricultural water withdrawals and industrial discharges can inform energy utilities about available resources, reducing risks of shortages or over-allocation.

Challenges to collaboration include differences in sectoral priorities, financial constraints, and governance gaps. Farmers may prioritize short-term yields over long-term conservation, while industries and energy producers may focus on profitability or regulatory compliance. Overcoming these barriers requires trust-building, clear agreements, and equitable distribution of costs and benefits. Strong facilitation by governments, NGOs, or basin authorities can help bridge these divides.

Collaboration with agriculture and energy sectors represents a practical and necessary pathway for industries seeking to improve water efficiency and resilience. By sharing resources, co-investing in infrastructure, and aligning goals, these sectors can collectively reduce water demand, secure reliable supplies, and protect ecosystems. As water stress intensifies under climate change, cross-sector collaboration will be a cornerstone of sustainable water management, ensuring that industry, agriculture, and energy systems thrive together rather than compete for limited resources.

Shared Infrastructure and Industrial Parks

Shared infrastructure and industrial parks are powerful models for advancing water conservation and efficiency in industry. By pooling resources and coordinating water management strategies, companies located within the same geographic area can achieve greater efficiencies than they would individually. These collaborative approaches not only reduce costs and resource use but also strengthen resilience, improve compliance with environmental regulations, and foster innovation in sustainable industrial development.

Industrial parks are purpose-built clusters where multiple businesses share physical infrastructure such as water supply systems, wastewater treatment facilities, and recycling units. Instead of each company building its own facilities, shared systems distribute costs, streamline operations, and enable more sophisticated technologies to be deployed at scale. For example, a centralized wastewater treatment plant serving an industrial park can incorporate advanced treatment technologies that might be prohibitively expensive for a single company. By consolidating demand and resources, industrial parks make water-efficient solutions more accessible and cost-effective.

Shared infrastructure also creates opportunities for industrial symbiosis, where one company's waste stream becomes another's resource. In water terms, this often involves the reuse of treated wastewater across different facilities. For instance, effluent from a food processing plant may be treated and reused for cooling by a neighboring power generator. Similarly, water used in textile dyeing could be treated and repurposed for irrigation or cleaning within the park. These exchanges reduce withdrawals from freshwater sources while minimizing discharges into the environment, aligning operations with circular economy principles.

Efficiency is further enhanced when water supply, treatment, and recycling systems are designed with scalability and flexibility in mind. Modular treatment units can be added as demand grows, while smart monitoring technologies allow operators to optimize flows in real time. By integrating digital solutions—such as sensors, data platforms, and predictive analytics—industrial parks can balance water use across multiple facilities, identify inefficiencies, and quickly respond to disruptions. This level of coordination would be difficult to achieve if each company managed its water resources independently.

Economic benefits are a major driver of shared infrastructure. Centralized water and wastewater systems reduce capital expenditures for individual companies, as the costs of construction and operation are distributed across multiple tenants. Operating expenses are also lowered by economies of scale, with larger treatment plants often operating more efficiently than smaller, standalone facilities. Access to reliable and cost-effective water infrastructure can also make industrial parks more attractive to investors and tenants, supporting local economic development while ensuring sustainable resource use.

Shared infrastructure in industrial parks also supports compliance with environmental regulations. Centralized treatment facilities can be designed to meet or exceed regulatory standards, reducing the risk of non-compliance for individual companies. This collective approach also makes it easier to monitor discharges, report performance, and engage with regulators. By demonstrating strong environmental performance at the park level, industries can build trust with surrounding communities and reduce the likelihood of conflicts over water use or pollution.

Governance and management structures are critical to the success of shared infrastructure. Clear agreements are needed to define responsibilities, allocate costs, and establish performance standards for participating companies. Park authorities, public-private partnerships, or third-party operators often play a central role in managing systems, ensuring fairness, and resolving disputes. Transparent governance creates confidence among tenants and fosters long-term cooperation, which is essential for maximizing the benefits of shared infrastructure.

Challenges remain, particularly in contexts where companies are hesitant to relinquish control over their water systems or where governance structures are weak. Upfront investment requirements can also pose barriers, especially in developing regions. However, international examples show that with strong planning, supportive policies, and effective stakeholder engagement, shared infrastructure can deliver substantial environmental and economic gains. Incentives such as tax breaks, financing support, or streamlined permitting can further encourage participation.

Shared infrastructure and industrial parks illustrate the potential of collective action in achieving water efficiency. By pooling resources, enabling symbiotic exchanges, and leveraging advanced technologies, these models reduce freshwater demand, lower costs, and enhance resilience. As industries confront growing water stress and stricter sustainability expectations, the adoption of shared systems in industrial clusters will become an increasingly important strategy for sustainable growth. Through collaboration and integrated planning, industrial parks can become hubs of innovation and leadership in water management.

Regional Partnerships for Water Security

Regional partnerships for water security are increasingly recognized as essential for addressing the complex challenges of managing shared resources. Water systems do not follow administrative boundaries, and the pressures of population growth, industrial expansion, climate change, and ecosystem degradation often extend beyond the jurisdiction of individual companies or governments. By forming partnerships at the regional scale, stakeholders can coordinate actions, share resources, and align goals to achieve more sustainable and equitable water outcomes.

These partnerships typically involve a diverse set of actors: governments, industries, agricultural producers, utilities, civil society organizations, and local communities. Each brings unique perspectives, resources, and responsibilities, making collaboration both challenging and valuable. Governments provide regulatory frameworks and policy alignment, industries contribute technical expertise and investment, while communities ensure that solutions are socially acceptable and equitable. The success of regional partnerships depends on building trust, creating transparent decision-making structures, and ensuring that the voices of all water users are heard.

One of the main advantages of regional partnerships is the ability to address water security at the basin or watershed level. Since rivers, aquifers, and ecosystems often serve multiple users, managing them collectively is the only way to balance competing demands and avoid conflicts. For example, a regional partnership can coordinate industrial water efficiency projects with agricultural irrigation improvements, ensuring that both sectors reduce demand without undermining each other. Such integrated approaches also create opportunities for shared infrastructure, such as multipurpose reservoirs or wastewater treatment facilities that serve multiple stakeholders.

Regional partnerships also foster resilience to climate change. Droughts, floods, and shifts in seasonal water availability require coordinated responses that no single actor can manage alone. By pooling data, forecasting tools, and emergency resources, regional coalitions can improve preparedness and reduce the impacts of extreme events. For industries, participating in such efforts helps safeguard supply chains, protect operations, and secure a social license to operate, while communities benefit from reduced vulnerability and improved services.

Financial collaboration is another strength of regional partnerships. Large-scale water security initiatives often require investments that exceed the capacity of individual actors. By forming partnerships, industries, governments, and financial institutions can co-invest in solutions such as desalination plants, advanced treatment systems, or ecosystem restoration projects. Joint financing arrangements not only distribute costs but also increase access to international funding sources, including development banks and climate finance mechanisms, which often prioritize collaborative, multi-stakeholder initiatives.

Governance structures are critical to the success of these partnerships. Effective regional coalitions establish clear rules for decision-making, cost-sharing, and accountability. They may create dedicated organizations or basin councils with representation from all stakeholder groups, supported by technical committees to provide scientific advice. Transparent governance helps prevent conflicts, ensures compliance with agreements, and builds long-term trust among participants.

Technology and data-sharing play a key enabling role. Regional partnerships often rely on shared monitoring systems, digital platforms, and hydrological models to inform decisions. By pooling information on water availability, consumption, and quality, stakeholders gain a clearer understanding of cumulative impacts and opportunities for efficiency. This transparency also reduces uncertainty and supports evidence-based decision-making.

Challenges to regional partnerships include balancing diverse interests, addressing power asymmetries, and overcoming institutional barriers. Industries may prioritize economic efficiency, while communities emphasize equity and ecosystems require environmental safeguards. Successful partnerships navigate these tensions through inclusive dialogue, adaptive management, and mechanisms that ensure benefits are distributed fairly. Supportive policies and legal frameworks are also necessary to align local, national, and regional priorities.

Regional partnerships for water security exemplify the principle that water challenges are shared and solutions must be collective. By coordinating across sectors and jurisdictions, these partnerships create synergies that reduce risks, lower costs, and build resilience for all stakeholders. As water scarcity intensifies globally, regional collaboration offers industries and communities alike a pathway to secure reliable water supplies, protect ecosystems, and promote sustainable development at scale.

Chapter 8: Financing Industrial Water Efficiency

Financing industrial water efficiency requires creative approaches that combine traditional investment mechanisms with emerging sustainable finance instruments. For many companies, the upfront costs of advanced treatment systems, recycling technologies, and digital monitoring tools can appear prohibitive, even when long-term savings are clear. Access to appropriate finance is therefore essential to accelerate adoption and scaling of water-efficient solutions. This chapter examines the spectrum of financing models available to industries, including conventional loans, green bonds, sustainability-linked loans, public-private partnerships, and blended finance structures. It highlights how aligning financial strategies with efficiency goals ensures that industries can secure the capital necessary to invest in resilient, sustainable water management practices.

Traditional and Innovative Financing Models

Financing is central to enabling industries to adopt water conservation and efficiency measures. While the technical solutions—ranging from advanced treatment technologies to closed-loop systems—are well established, many companies struggle with the upfront costs of implementation. Both traditional and innovative financing models play critical roles in bridging this gap, ensuring that industries can invest in sustainable water management without jeopardizing competitiveness.

Conventional approaches remain widely used and form the foundation of industrial water investments. Bank loans are one of the most common instruments, allowing companies to secure capital for infrastructure upgrades such as wastewater treatment plants, smart metering systems, or recycling technologies. Loans typically rely on the creditworthiness of the borrower and may be supported by collateral. While effective, they can be less accessible to small and medium-sized enterprises (SMEs) that lack strong credit histories.

Corporate capital expenditure (CAPEX) allocations are another traditional method. Many industries finance water efficiency directly from internal budgets, particularly when the payback period is short and efficiency measures deliver measurable cost savings. These investments are often justified through reduced operating expenses, lower energy use, and decreased water procurement costs. However, CAPEX budgets are frequently constrained, and water projects may compete with other business priorities for limited funds.

Government subsidies, grants, and low-interest financing programs also play a role. In regions where water scarcity poses significant risks, governments often provide financial incentives to encourage adoption of water-efficient technologies. These incentives may cover part of the capital cost or provide tax relief for qualifying projects. While helpful, reliance on subsidies can be uncertain due to shifting policy priorities and budget constraints.

To overcome the limitations of traditional methods, new and more flexible financing approaches have emerged. One such model is performance-based contracting, often implemented through energy service companies (ESCOs) or water service companies (WASCOs). Under this arrangement, a third-party provider finances, installs, and manages water efficiency upgrades, with repayment tied to the savings achieved. This model reduces upfront costs for industries and transfers some of the performance risk to the provider.

Green bonds and sustainability-linked loans are increasingly used to finance water efficiency projects. Green bonds raise capital specifically for environmentally beneficial projects, while sustainability-linked loans tie interest rates to the borrower's achievement of defined sustainability targets, including water conservation. These instruments not only provide access to capital but also enhance corporate reputation by demonstrating alignment with ESG goals.

Public-private partnerships (PPPs) represent another innovative model, particularly for large-scale water infrastructure projects. In this arrangement, governments and industries share investment responsibilities and risks. For example, a government may build a centralized treatment facility while industries commit to purchasing treated water or reusing effluent. PPPs are particularly valuable in regions where both public and private actors face resource constraints but share an interest in securing long-term water supplies.

Blended finance has gained traction as a way to mobilize private investment using concessional funds from development banks, donors, or philanthropic organizations. By lowering the risk profile of water projects, blended finance attracts private capital that might otherwise avoid such investments. This model is especially important in developing economies, where industries face high water risks but limited access to affordable financing.

The effectiveness of both traditional and innovative financing models depends on enabling conditions such as clear regulatory frameworks, supportive policies, and reliable data on water use and savings. Transparent reporting on water performance helps investors assess risk and return, while strong governance reduces uncertainty. Digital monitoring and smart metering can strengthen the financial case by providing verifiable data on efficiency gains, ensuring that savings are measurable and bankable.

Despite the growing range of financing models, barriers remain. Small industries often lack access to capital or technical expertise to navigate financing options. Investors may perceive water projects as high risk due to uncertainties in regulatory enforcement, climate impacts, or return timelines. Overcoming these challenges requires stronger collaboration between industries, financiers, and policymakers. Expanding blended finance mechanisms, standardizing reporting frameworks, and promoting knowledge-sharing platforms can help bridge gaps and accelerate investment.

Traditional and innovative financing models together create a spectrum of options that can be tailored to industry needs. While traditional loans and CAPEX remain important, innovative approaches like performance-based contracts, green bonds, and blended finance are opening new pathways for investment. By leveraging this range of tools, industries can secure the resources needed to adopt water-efficient technologies, strengthen resilience, and contribute to broader sustainability goals.

Public-Private Partnerships

PPPs have become a cornerstone of advancing industrial water conservation and efficiency, particularly in contexts where large-scale investments and shared responsibilities are required. These arrangements bring together governments, private sector actors, and sometimes civil society organizations to co-design, finance, build, and operate water-related infrastructure or services. By leveraging the strengths of each partner, PPPs enable solutions that individual actors would struggle to implement alone.

One of the most significant advantages of PPPs is the pooling of resources. Governments often face budgetary constraints that limit their ability to invest in advanced water infrastructure, while private companies may be reluctant to commit capital to projects with long payback periods. Through PPPs, governments provide regulatory support, guarantees, or subsidies, while the private sector contributes technical expertise, innovation, and financing. This shared responsibility allows large-scale projects—such as desalination plants, wastewater treatment facilities, or industrial recycling systems—to move forward more quickly and efficiently.

PPPs are also valuable in aligning public goals with private incentives. Governments are tasked with ensuring water security, environmental protection, and equitable access, while industries seek reliable, affordable water supplies and operational continuity. A well-designed PPP can reconcile these objectives by establishing clear performance targets for efficiency, sustainability, and compliance. For example, private partners may be contracted to deliver a certain volume of recycled water or maintain systems that reduce losses, with compensation tied to measurable outcomes. This performance-based approach ensures accountability while creating mutual benefits.

Shared infrastructure is a frequent focus of PPPs. Instead of each company building and managing its own treatment facility, a centralized plant operated under a PPP can serve multiple industrial tenants, municipalities, or agricultural users. This model reduces duplication of investments, achieves economies of scale, and enables the deployment of advanced technologies that smaller entities could not afford independently. By consolidating services under a PPP framework, stakeholders also benefit from more consistent water quality, improved monitoring, and reduced environmental impacts.

Beyond infrastructure, PPPs play a role in advancing innovation and capacity building. Private partners often bring cutting-edge technologies and operational expertise that governments or public utilities may lack. In turn, governments provide the regulatory certainty and long-term contracts needed to encourage private investment. Training programs, knowledge exchange, and technology transfer embedded in PPPs strengthen local capacity, ensuring that efficiency gains are sustained over time.

Financial innovation is another benefit. PPPs can unlock blended finance by combining public funds, private capital, and contributions from development banks or donors. This reduces risks for investors while expanding access to financing for projects that might otherwise be considered too costly or uncertain. For industries, participating in such partnerships spreads costs across multiple stakeholders, making water efficiency investments more financially viable.

Despite these advantages, PPPs face challenges that require careful management. Misaligned incentives, unclear contractual terms, or inadequate regulation can undermine performance and trust.

Communities may also perceive PPPs as prioritizing private profit over public benefit, particularly if tariffs rise or access is restricted. Addressing these concerns requires transparency, strong governance frameworks, and mechanisms for stakeholder participation. Contracts must clearly define responsibilities, performance indicators, and dispute resolution processes to ensure accountability and fairness.

Another challenge is balancing long-term sustainability with short-term financial goals. While PPPs can deliver immediate efficiency improvements, they must also ensure that operations remain sustainable over decades. This includes accounting for climate risks, changing water availability, and evolving regulatory standards. Adaptive management provisions in PPP contracts help address these uncertainties by allowing flexibility to adjust strategies as conditions change.

Public-private partnerships thus represent a powerful tool for scaling up industrial water conservation and efficiency. By combining the resources and expertise of governments and the private sector, PPPs enable infrastructure development, innovation, and financial solutions that would otherwise be out of reach. When designed transparently and managed effectively, they balance economic, environmental, and social goals, contributing to long-term water security and resilience. In a future marked by increasing water stress, PPPs will continue to play a central role in aligning industrial growth with sustainable resource management.

Role of Green Bonds and Sustainability-Linked Loans

Green bonds and sustainability-linked loans are emerging as central financing tools for advancing industrial water efficiency. They provide a mechanism for companies to align financial performance with sustainability outcomes while accessing growing pools of capital tied to environmental, social, and governance priorities. With investors increasingly concerned about water scarcity, pollution, and regulatory risks, these instruments enable industries to demonstrate leadership and credibility in their water management strategies.

Green bonds are specifically designed to raise funds for environmentally beneficial projects, with proceeds earmarked for uses such as advanced wastewater treatment, closed-loop cooling, recycling systems, or smart monitoring infrastructure. Issuers are expected to align with internationally recognized frameworks like the Green Bond Principles, which require clear definitions of eligible projects, transparent selection processes, and regular reporting. For water-related projects, performance metrics might include reductions in withdrawals, higher recycling rates, or improvements in effluent quality. This level of accountability builds investor trust while helping industries showcase concrete progress. Green bonds also improve corporate reputation, expand access to sustainability-oriented investors, and in some cases lower borrowing costs due to strong market demand.

Sustainability-linked loans differ in structure by tying the cost of borrowing to the borrower's performance on specific sustainability targets rather than restricting use of proceeds. Targets often include water intensity reduction across operations, implementation of water reuse systems, or achieving recognized stewardship certifications. If the borrower meets or exceeds these targets, interest rates are reduced; if not, costs rise. This model creates financial incentives for continuous improvement and embeds accountability directly into loan agreements. Because funds can be used flexibly, companies can direct capital toward the areas of greatest need while still holding themselves accountable to measurable water outcomes.

Both instruments reinforce transparency through external verification and public reporting. Independent reviewers often confirm that projects financed through green bonds meet stated objectives, while sustainability-linked loan targets must be credible and verifiable. This not only reassures investors but also ensures that companies maintain focus on achieving genuine improvements. Integration with disclosure frameworks such as CDP Water Security or GRI further strengthens alignment between corporate water management and ESG reporting expectations.

Adoption of these financing tools delivers both financial and strategic benefits. Access to capital improves, particularly from investors with sustainability mandates, while reduced borrowing costs can make efficiency projects more economically attractive. Publicly linking financial performance to water outcomes enhances credibility with stakeholders, strengthens community trust, and secures a more robust social license to operate. For industries in water-stressed regions, these advantages can be critical to long-term viability.

Challenges remain in applying these models. Setting ambitious yet realistic water-related targets can be complex, especially in industries with diverse global operations or variable water demands. Reporting and verification requirements may increase administrative burdens, and smaller firms may lack the capacity to issue bonds or negotiate sustainability-linked terms. However, supportive frameworks, blended finance solutions, and growing market maturity are helping to lower these barriers.

The role of green bonds and sustainability-linked loans is likely to expand as sustainable finance becomes more mainstream. They not only mobilize capital for specific projects but also integrate water management into the broader financial architecture of companies. By linking access to finance with water efficiency, these instruments make sustainability a core driver of business resilience and competitiveness. In an increasingly water-constrained world, they represent a powerful means of aligning industrial growth with responsible stewardship of critical resources.

Economic Valuation of Water Efficiency Benefits

The economic valuation of water efficiency benefits is essential for industries to understand the full return on investment in conservation measures. While water is often undervalued in monetary terms, its role in sustaining operations, enabling compliance, and mitigating risks makes it a critical resource. By quantifying the direct and indirect benefits of efficiency improvements, companies can justify investments, attract financing, and build stronger cases for integrating water management into long-term business strategies.

Direct financial benefits are the most immediate and measurable outcomes of water efficiency. Reducing withdrawals lowers utility bills or pumping costs, while decreasing wastewater volumes cuts treatment and disposal expenses. In many industries, energy savings are another significant gain, as heating, cooling, and pumping less water reduces electricity or fuel use. These direct savings often provide relatively short payback periods for investments in technologies such as smart metering, recycling systems, or high-efficiency equipment. For companies facing rising tariffs or operating in water-stressed regions, the financial case becomes even stronger as the cost of inaction escalates.

Indirect benefits extend beyond operational savings and are equally important to capture in valuations. Efficient water use reduces exposure to regulatory penalties or compliance costs, as companies can more easily meet discharge and withdrawal limits. It also strengthens resilience to scarcity, minimizing the risk of production disruptions during droughts or supply restrictions. Avoided downtime can translate into substantial financial savings, particularly in sectors such as power generation, food processing, or electronics manufacturing where continuous operations are critical. These indirect benefits often outweigh the upfront costs of efficiency investments when considered over the medium to long term.

Another dimension of valuation is reputational and market benefit. Stakeholders—including investors, consumers, and communities—are increasingly scrutinizing corporate water practices. Companies that demonstrate leadership in water efficiency can enhance brand value, differentiate themselves from competitors, and secure stronger social licenses to operate. In many markets, sustainability credentials directly influence purchasing decisions, while in capital markets, investors reward companies with lower risk profiles and stronger environmental performance. Quantifying these benefits can be challenging, but case studies show that improved reputation can lead to higher sales, stronger investor confidence, and reduced costs of capital.

Economic valuation must also consider risk reduction. Water-related risks—such as scarcity, flooding, or pollution—pose material threats to profitability and continuity. Efficiency investments act as insurance against these risks by reducing dependence on vulnerable sources and strengthening adaptive capacity. Assigning monetary values to avoided risks, such as the costs of production stoppages or reputational damage from pollution incidents, provides a fuller picture of the benefits of efficiency. Tools such as scenario analysis and probabilistic modeling can help quantify these avoided costs, making them part of financial decision-making.

At a broader level, valuation should incorporate ecosystem and societal benefits. Reduced withdrawals ease pressure on local water systems, helping sustain ecosystems and communities that industries depend on indirectly. These benefits may not always be reflected in balance sheets but can influence long-term operational security and stakeholder relationships. For example, industries that conserve water contribute to healthier aquifers, rivers, and wetlands, reducing conflicts with other users and ensuring reliable supplies in the future. Monetizing these benefits through ecosystem service valuation frameworks can demonstrate how water efficiency creates shared value beyond the factory gate.

The challenge in valuing water efficiency lies in data availability and methodological complexity. While direct savings are straightforward to calculate, indirect and societal benefits often require assumptions and modeling. However, advances in digital monitoring, life cycle assessment, and risk valuation tools are making it easier to quantify water's true value. Integrating these methods into cost-benefit analyses allows companies to build more compelling cases for efficiency projects and to prioritize investments with the greatest overall impact.

By adopting a comprehensive approach to economic valuation, industries can move beyond seeing water efficiency as a cost to recognizing it as a driver of financial resilience, competitiveness, and sustainability. Quantifying both tangible and intangible benefits provides the evidence needed to secure financing, attract investors, and justify strategic investments. In an era of intensifying water scarcity and climate uncertainty, placing a robust economic value on water efficiency is not only prudent but essential for long-term industrial success.

Chapter 9: Future Pathways for Industrial Water Use

The future of industrial water use will be defined by the ability of companies to innovate, adapt, and align with broader sustainability agendas. As pressures from climate change, population growth, and competing demands intensify, industries will need to move beyond incremental efficiency gains toward transformative approaches that reshape how water is valued and managed. This chapter explores potential pathways for industries to achieve resilience, including the adoption of emerging technologies, integration into circular economy systems, stronger governance frameworks, and alignment with global policy goals. By anticipating these pathways, industries can position themselves to thrive in an era where water is both a critical resource and a central driver of sustainable development.

Emerging Technologies and Digital Solutions

Emerging technologies and digital solutions are transforming industrial water management, offering unprecedented opportunities for efficiency, resilience, and sustainability. Where traditional methods relied heavily on manual monitoring and reactive maintenance, new tools now allow industries to measure, predict, and optimize water use in real time. By integrating digital platforms with advanced treatment and reuse systems, companies can shift from linear consumption models toward smarter, more adaptive, and circular approaches. These innovations not only conserve water but also strengthen competitiveness and reduce operational risks in a resource-constrained world.

At the core of this transformation is the Internet of Things (IoT). Networks of sensors installed throughout industrial facilities continuously track variables such as flow rate, pressure, temperature, and water quality parameters like pH, turbidity, and salinity. The data generated allows companies to pinpoint inefficiencies, identify leaks instantly, and ensure compliance with regulatory standards. Unlike periodic manual checks, IoT provides continuous visibility, enabling real-time decision-making. For industries operating at large scales or across multiple sites, this level of monitoring is invaluable for standardizing performance and identifying hotspots of water use.

AI and machine learning build on these capabilities by analyzing the vast amounts of data generated by sensors. Predictive algorithms can forecast water demand based on production schedules, seasonal variations, or external factors such as weather patterns. They can also anticipate equipment failures, scaling, or fouling in treatment systems, enabling proactive maintenance that reduces downtime and costs. Over time, AI learns from patterns in the data, providing increasingly precise recommendations for optimizing water use, recycling, and discharge management.

Digital twins are another groundbreaking application in industrial water management. A digital twin is a virtual model of a facility's water systems that mirrors real-world conditions using live data from sensors. Operators can use these models to simulate different scenarios, such as droughts, equipment upgrades, or process changes, and assess their impacts on water use and efficiency. Digital twins provide a safe, cost-effective environment for testing strategies before implementing them in practice. They are especially valuable in industries with complex water systems, such as energy generation or chemical manufacturing, where even small inefficiencies can translate into significant costs.

Blockchain technology is emerging as a solution for transparency and accountability, particularly in supply chain water management. By creating secure, tamper-proof records of water use, quality, and transactions, blockchain ensures that data shared between suppliers, regulators, and customers is trustworthy. This is particularly relevant for companies aiming to demonstrate compliance with ESG reporting requirements or to assure investors that sustainability targets are being met. Blockchain can also facilitate water trading schemes, enabling industries to buy, sell, or lease water rights in a transparent marketplace.

In treatment and recycling, advanced membrane technologies, electrochemical systems, and nanomaterials are enabling higher efficiency and lower energy use. For instance, nanofiltration membranes with improved resistance to fouling extend the lifespan of systems and reduce maintenance costs. Electrochemical oxidation technologies can break down contaminants that traditional methods struggle to remove, producing water clean enough for reuse. Pairing these innovations with digital monitoring ensures that treatment systems operate optimally, adapting to fluctuations in wastewater quality and demand.

Cloud-based platforms are increasingly being used to integrate data from diverse sources, creating centralized dashboards for water management. These platforms allow managers to oversee performance across multiple facilities, benchmark results, and identify best practices. Remote accessibility is particularly valuable for multinational companies or industrial parks where centralized oversight can drive consistency. Coupled with mobile applications, these platforms empower operators and managers to respond quickly to issues, even when off-site.

The integration of renewable energy with digital water solutions represents another important frontier. Solar-powered pumping systems combined with smart controls reduce both energy and water footprints. AI-optimized cooling systems in energy plants can balance efficiency with minimal water use, while predictive analytics ensure operations remain resilient under changing climate conditions. These synergies demonstrate how digital and technological innovations in water management align closely with decarbonization goals.

Despite the promise, challenges to adoption remain. High upfront costs, cybersecurity concerns, and the need for skilled personnel can be barriers, particularly for small and medium-sized enterprises. Data integration across legacy systems is also complex, requiring careful planning and investment. However, as costs decline and solutions become more scalable, these barriers are gradually diminishing. Supportive policy frameworks, capacity-building initiatives, and collaborative platforms can further accelerate adoption.

Emerging technologies and digital solutions are reshaping industrial water management from the ground up. By enabling real-time monitoring, predictive optimization, transparent reporting, and advanced treatment, they allow companies to achieve efficiencies previously unattainable with traditional methods. More importantly, they provide the tools to build resilience against climate change, regulatory pressures, and market expectations. As industries move toward circular and sustainable models, these innovations will serve as the backbone of water stewardship, ensuring that efficiency is not only a technical achievement but also a strategic driver of long-term success.

Scenario Planning for Water Scarcity

Scenario planning for water scarcity is a powerful tool that allows industries to anticipate potential futures and prepare strategies to remain resilient under uncertain conditions. Unlike traditional forecasting, which projects trends in a linear fashion, scenario planning develops multiple plausible futures shaped by variables such as climate change, population growth, regulatory shifts, and technological innovation. By considering these different pathways, industries can identify vulnerabilities, test the robustness of their strategies, and invest in solutions that perform well across a range of possible outcomes.

The first step in scenario planning is identifying the drivers of water scarcity that are most relevant to an industry. These include external factors such as climate variability, changes in rainfall patterns, groundwater depletion, and competing demands from agriculture and municipalities. Internal drivers, like production processes, supply chain dependencies, and water efficiency practices, also shape a company's exposure to scarcity risks. Mapping these drivers creates a foundation for exploring how combinations of external and internal pressures may evolve into different futures.

Industries then construct scenarios that represent plausible conditions. One scenario might envision intensified droughts leading to higher water tariffs and stricter regulations. Another could depict technological breakthroughs that expand access to recycled or desalinated water, reducing scarcity risks. A third may focus on societal pressures, where community opposition to industrial withdrawals forces industries to cut consumption regardless of availability. Each scenario highlights different challenges and opportunities, ensuring that companies consider a full range of possible futures rather than relying on a single projection.

Quantitative and qualitative analysis both play important roles in scenario planning. Quantitative methods, such as water demand modeling, climate impact projections, and financial simulations, provide numerical estimates of risks and outcomes. Qualitative methods, such as expert interviews and stakeholder workshops, capture insights into political, cultural, and behavioral factors that shape water availability. Combining the two approaches enriches the scenarios, ensuring they are both data-driven and grounded in the broader social context.

Industries benefit from using scenario planning to stress-test their strategies. For example, a company considering investment in a new water-intensive facility can test how it would perform under scenarios of rising scarcity, higher energy costs, or stricter regulations. If the facility proves resilient across multiple scenarios, the investment is more secure. If not, the company may opt for alternatives such as modular designs, water-efficient technologies, or diversifying production to less water-stressed regions. Scenario planning also guides contingency planning, helping companies prepare response strategies for sudden shocks such as extreme droughts or supply disruptions.

Another advantage of scenario planning is its ability to foster innovation. By envisioning futures where water is scarce and expensive, companies are motivated to develop solutions that minimize dependence on freshwater resources. This can accelerate adoption of closed-loop systems, digital monitoring technologies, and partnerships with other sectors to share water infrastructure. Scenario planning can also identify opportunities for collaboration with policymakers, communities, and other stakeholders to develop basin-level solutions that benefit all users.

Challenges in implementing scenario planning include the uncertainty inherent in projecting long-term futures, the complexity of integrating diverse data sources, and the need for organizational commitment. Industries may struggle to allocate resources to scenario planning when short-term financial pressures dominate decision-making. However, the value of scenario planning lies precisely in its ability to prepare companies for long-term risks that could undermine profitability and resilience. When integrated into enterprise risk management and strategic planning, scenario planning strengthens a company's capacity to adapt to uncertain and volatile conditions.

Scenario planning for water scarcity provides industries with a structured way to navigate uncertainty and build resilience. By exploring multiple futures shaped by environmental, regulatory, and societal drivers, companies can identify risks, test strategies, and uncover opportunities for innovation. This proactive approach shifts water management from reactive crisis response to forward-looking resilience building, ensuring that industries are prepared not only to survive but also to thrive in an era of increasing water stress.

Policy Directions for Industrial Water Resilience

Policy directions for industrial water resilience are becoming increasingly critical as industries face growing risks from scarcity, climate change, and rising stakeholder expectations. Governments, regulators, and industry leaders are recognizing that fragmented approaches are insufficient to safeguard resources, and that coherent, forward-looking policy frameworks are necessary to balance economic growth with sustainable water use. By setting clear rules, providing incentives, and fostering collaboration, policy plays a central role in enabling industries to adapt and thrive under conditions of increasing water stress.

One key policy direction is the establishment of comprehensive water pricing and tariff structures that reflect the true cost of water. Historically, underpriced water has discouraged efficiency, making it cheaper for industries to overuse rather than conserve. By incorporating the costs of treatment, distribution, scarcity, and environmental degradation into pricing frameworks, governments create financial signals that incentivize conservation and investment in efficiency technologies. To ensure fairness, policies can include tiered structures that encourage efficient use while safeguarding affordability for essential needs.

Another important policy direction lies in strengthening regulatory frameworks for water quality and discharge. Clear, enforceable standards on wastewater treatment, pollution prevention, and reuse create the regulatory certainty industries need to invest in compliance technologies. Forward-looking policies also promote the adoption of advanced treatment systems and water recycling, moving beyond minimum compliance toward circular water use. This not only reduces environmental impacts but also enhances resilience by expanding nontraditional sources of supply.

Policies promoting industrial water audits and disclosure represent a growing trend. By mandating regular audits and requiring companies to report on water use, risks, and efficiency measures, governments create transparency that drives accountability. Public reporting aligns with environmental, social, and governance expectations, allowing stakeholders to track progress and apply pressure where needed. Incentives for voluntary disclosure, combined with requirements for high-risk sectors, encourage industries to adopt best practices in monitoring and reporting.

Collaboration-focused policies are equally vital. Basin-level governance frameworks that integrate industrial, agricultural, and municipal users enable collective management of shared resources.

Governments can facilitate the formation of water stewardship partnerships, industrial parks with shared infrastructure, and regional coalitions that coordinate demand management. By embedding industry participation in basin councils or watershed committees, policies ensure that companies contribute to collective solutions rather than pursuing fragmented approaches that may undermine overall resilience.

Financial and innovation policies also play an enabling role. Tax incentives, subsidies, and preferential financing for efficiency projects lower the barriers for industries to adopt advanced technologies. Policies promoting green bonds, sustainability-linked loans, and blended finance mechanisms channel capital toward water-resilient infrastructure. Similarly, investment in research and development accelerates the commercialization of emerging technologies, ensuring industries can access innovative solutions at scale.

Climate adaptation policies are increasingly being aligned with water resilience objectives. National adaptation plans and sectoral strategies often identify industrial water efficiency as a critical component of resilience. Policies that integrate water into broader climate frameworks encourage industries to account for long-term risks, such as altered rainfall patterns or extreme droughts, in their planning and operations. This alignment strengthens the policy coherence needed to address interconnected challenges across water, energy, and climate.

Enforcement and compliance mechanisms remain a cornerstone of effective policy. Regulations without monitoring and penalties risk being ineffective. Policies must therefore ensure that enforcement agencies are adequately resourced and that penalties for noncompliance are meaningful. At the same time, supportive policies should reward companies that go beyond compliance, creating recognition programs or preferential access to permits for leaders in efficiency and conservation.

Finally, international cooperation influences national and industrial policy directions. Frameworks such as the UN Sustainable Development Goals and agreements on transboundary water management encourage countries to adopt policies that harmonize with global sustainability agendas. Industries operating internationally benefit from this alignment, as consistent standards across jurisdictions reduce complexity and create a level playing field.

Policy directions for industrial water resilience thus encompass a blend of regulatory certainty, financial incentives, collaborative governance, and integration with broader sustainability goals. Together, these approaches provide the enabling environment industries need to reduce risks, invest in innovation, and contribute to long-term water security. As pressures on water systems intensify, policy will remain a decisive factor in shaping how industries adapt, compete, and sustain themselves in an era of increasing scarcity.

Scaling Innovations Globally

Scaling innovations globally in industrial water conservation and efficiency is essential to address the shared challenges of scarcity, climate change, and rising demand across industries. While pilot projects and localized solutions demonstrate the technical feasibility of advanced water management practices, their broader impact depends on the ability to replicate and adapt them across regions, sectors, and markets. Scaling requires more than simply transferring technology; it involves creating enabling conditions, building capacity, and aligning policies and incentives to foster widespread adoption.

A primary driver of global scaling is knowledge transfer. Many industries are already deploying cutting-edge technologies such as smart metering, digital twins, and advanced recycling systems in water-stressed regions. Sharing these successes through global platforms, industry associations, and multilateral initiatives enables other companies and countries to learn and replicate. Open-access knowledge hubs, best-practice guidelines, and international benchmarking frameworks accelerate this diffusion by providing industries with tested approaches that can be tailored to local contexts.

Technology providers and multinational corporations play an important role in scaling innovations. Companies with global operations can implement proven solutions across multiple facilities, adjusting designs to local regulatory and resource conditions. This creates internal economies of scale, lowers costs, and builds a track record of successful implementation. Technology providers, meanwhile, can expand global markets by standardizing products and offering flexible solutions that meet diverse industrial needs. Partnerships between global firms and local enterprises further support adoption, combining technical expertise with contextual understanding.

Policy frameworks also strongly influence the scaling of innovations. Countries that align regulatory standards, provide incentives, and support international cooperation create fertile ground for industries to adopt new practices. Trade agreements, climate compacts, and shared sustainability commitments encourage the transfer of technologies across borders. International organizations such as the United Nations, World Bank, and regional development banks promote scaling by funding large-scale projects, supporting pilot-to-market transitions, and facilitating south-south cooperation.

Finance is another critical enabler of scaling. The upfront costs of water-efficient technologies can deter adoption, particularly in emerging markets. Access to international funding through mechanisms such as green bonds, blended finance, and climate funds reduces this barrier by distributing risks and offering concessional terms. When industries demonstrate successful implementation, financial institutions are more likely to support similar projects elsewhere, creating a reinforcing cycle of investment and adoption.

Cultural and institutional factors must also be addressed in scaling efforts. Innovations that succeed in one region may encounter resistance in another due to different social norms, governance systems, or industrial practices. Effective scaling therefore requires adaptation to local conditions rather than one-size-fits-all replication. Capacity building, including training programs, technical support, and education initiatives, ensures that industries and communities can operate and maintain new technologies effectively. Collaborative partnerships with local stakeholders are essential to building trust and ensuring solutions are socially acceptable.

Digitalization enhances global scaling by creating platforms that connect stakeholders and streamline knowledge exchange. Online monitoring systems allow industries to compare performance across regions, identify best practices, and track progress toward shared goals. Digital marketplaces enable technology providers to reach international buyers, while cloud-based tools facilitate real-time collaboration across borders. These digital networks reduce transaction costs and accelerate the spread of innovations to regions where they are most urgently needed.

Challenges to scaling remain, including uneven regulatory environments, disparities in financial resources, and varying levels of technical capacity. Intellectual property concerns and market fragmentation can also slow diffusion. Overcoming these barriers requires coordinated efforts between governments, industries, investors, and international organizations. Multistakeholder initiatives that combine technical expertise, financial support, and policy advocacy are among the most effective ways to address these obstacles.

Scaling innovations globally is not simply about replicating technology but about building systems that enable widespread and sustainable adoption. By aligning knowledge sharing, finance, policy, and local adaptation, industries can accelerate the transition toward efficient and resilient water use worldwide. In an interconnected global economy, where water challenges cross borders, scaling innovations ensures that progress in one region can spark solutions across many others, creating collective benefits for industry, communities, and ecosystems alike.

Conclusion

Industrial water conservation and efficiency are no longer optional practices but core strategies for resilience, competitiveness, and sustainability in a rapidly changing world. Water has traditionally been undervalued, treated as an abundant and low-cost input, yet the realities of climate change, growing populations, and shifting regulatory frameworks have made clear that industries cannot thrive without safeguarding this critical resource. The insights explored throughout this book highlight that water efficiency is not only about reducing withdrawals but about transforming the way industries operate, invest, and collaborate to align with a more resource-constrained future.

A central theme is that industries face both risk and opportunity in their relationship with water. Risks include rising operational costs, potential production disruptions, and reputational challenges as stakeholders increasingly demand transparency and accountability. At the same time, opportunities lie in innovation, where emerging technologies, advanced monitoring systems, and digital solutions offer unprecedented visibility and control over water use. By investing in these tools and adopting practices such as closed-loop systems, wastewater reuse, and integration with circular economy principles, industries can strengthen resilience while reducing costs and improving long-term competitiveness.

Policy frameworks, financing mechanisms, and governance structures are equally important in shaping the path forward. Effective policies provide clarity and incentives for conservation, while financial tools such as green bonds, sustainability-linked loans, and public-private partnerships unlock capital for scaling solutions. Corporate governance structures that integrate water into enterprise risk management and ESG reporting ensure that water efficiency is embedded within broader sustainability strategies. Together, these elements create the enabling environment that allows industries to move from fragmented actions toward systemic transformation.

Another major insight is the importance of collaboration. Industrial facilities do not exist in isolation; they operate within shared basins and communities where water is a common resource. Partnerships at regional levels, cooperation with agriculture and energy sectors, and participation in basin governance mechanisms are vital for balancing competing demands and building collective resilience. Sharing infrastructure, engaging in industrial symbiosis, and supporting community water security strengthen the social license to operate while contributing to broader development goals.

Scenario planning, global scaling of innovations, and continuous improvement provide industries with the tools to navigate uncertainty. The future of water availability is unpredictable, but companies that plan for multiple scenarios, adapt quickly, and invest in solutions that work across contexts will be better prepared for shocks. Scaling innovations globally ensures that best practices in one region can inform and inspire actions elsewhere, creating synergies that benefit industries, communities, and ecosystems alike.

Ultimately, the path to industrial water resilience requires a balance of innovation, policy support, financial commitment, and collective action. Efficiency and conservation must be seen not as constraints but as opportunities to unlock new value, reduce risks, and ensure long-term sustainability. By aligning operations with global sustainability agendas and embracing the principles of stewardship, industries can position themselves as leaders in addressing one of the defining challenges of the 21st century. The imperative is clear: conserving and using water wisely is fundamental not only to industrial success but to the resilience of societies and ecosystems worldwide.

www.ingramcontent.com/pod-product-compliance
Lightning Source LLC
Chambersburg PA
CBHW050530240926
51846CB00003B/767
9781991370990